IMAGES
of America

DEL MAR
FAIRGROUNDS

ON THE COVER: When La Jolla High School student Raquel Tejada was crowned Fairest of the Fair for the 1958 season, she became an ambassadress of the Del Mar Fairgrounds. As such, she participated in many promotions with ambassador Tommy Hernandez, who appeared during the fair in character as Don Diego. These often included photo shoots like this one, with Grandstand entertainment artists Mariachi Los Aventureros. They were known countywide for their lively Mexican-style music, which entertained fair patrons during several fairs. Pictured from left to right are Hernando Tellez, Al Arreda, Raquel Tejada Welch, Don Diego, Pat Hobbs, Frank "Pancho" Pena, (leader) Miguel Aranda, and Jose Mata. (Courtesy 22nd District Agricultural Association [DAA].)

IMAGES
of America

DEL MAR
FAIRGROUNDS

Diane Y. Welch, B. Paul Welch, and the
22nd District Agricultural Association

ARCADIA
PUBLISHING

ISBN 978-1-5316-3742-2

Published by Arcadia Publishing
Charleston SC, Chicago IL, Portsmouth NH, San Francisco CA

Library of Congress Catalog Card Number: 2007941840

For all general information contact Arcadia Publishing at:
Telephone 843-853-2070
Fax 843-853-0044
E-mail sales@arcadiapublishing.com
For customer service and orders:
Toll-Free 1-888-313-2665

Visit us on the Internet at www.arcadiapublishing.com

This book is dedicated to James E. Franks,
the father of the Del Mar Fairgrounds.

CONTENTS

ACKNOWLEDGMENTS

Our heartfelt thanks go to the staff of the 22nd DAA, without whose generosity, patience, and guidance this book would not have been made. The scope of its archives is so vast that literally thousands of images could not be included. As a result, we limited the Del Mar Fairgrounds' history with a few exceptions (such as a much needed chapter on today's fairground) to the 1930s, 1940s, and 1950s. It is hoped that a second volume of fairground history will spotlight later decades. It will certainly include the stars of the grandstand who are deserving of their recorded place in the fair's storied past.

Notably, the following have been indispensable: Joanne Gaines and Jimmie Marie English for their photographs and recollections of Jim Franks; veteran newsman and former fairground publicity man Bill Arballo—huge thanks for our book's lively introduction; the directors of the 22nd DAA and of the Del Mar Thoroughbred Club for allowing open access to their archives; Tim Fennell, Katie Phillips, Heidi Rheinholz, Jane Spivey, Kina Paegert, and Jill Coughlin of the 22nd DAA for their helping hands; Mac MacBride of the Del Mar Thoroughbred Club for incredible period photographs; Ron Bosley of the International Club Crosby for contributing some rare photographs of Crosby; Lisa Iaria of the Del Mar National Horse Show for its period photographs; Lucy Berk of the Escondido Library Pioneer Room for her research help; Susan Guttierez of the Carlsbad City Library History Room for Fred Mitchell's photograph; Mary Allely of the National City Library Morgan History Room for research help and the portrait of Frank Kimball; Don Terwilliger for his father's photograph; Robert Hernandez for his own rare photographs; photographer Tom Keck for his unique shots of the aftermath of the Heaven's Gate tragedy; and Debbie Seracini for guiding us diligently through this project. Last, but not least, we thank the incredible people who attended and took part in the Del Mar Fairgrounds over the decades. Their photographic images have been recorded here for posterity. We salute you friends. *Le saludamos amigos.*

INTRODUCTION

The San Diego County Fair has a storied past. Through several decades of uncertainty, it eventually evolved into the annual fair hosted at the Del Mar Fairgrounds that we know and love today.

In 1880, Frank Kimball spearheaded the first ever county fair in National City on a site now known as Kimball Park. It was dedicated to the Kimball family in appreciation of their many years of service to the community they founded. Frank Kimball would later serve as the state commissioner of horticulture. The fair was sponsored by the National Ranch Grange, which was made up of 50 fruit growers and farmers. It drew a larger than expected attendance on opening and throughout its run. Some visitors came by train from as far away as San Francisco. The fair continued in National City for several years and then was picked up by interested parties in Escondido in the late 1880s. Among the news accounts were stories about Wyatt Earp being the equestrian judge. Exhibits were housed in a circus tent erected at Washington and Grant Streets. Admission for grown ups was a quarter, and kids were admitted for 15¢.

Horse racing on the beach with side betting was a major attraction when the fair was held in Coronado in the Armory Hall at Fourth and Ash Streets. Julian was reported to be a major exhibitor of fruits, grain, and minerals. It took all the blue ribbons in the deciduous fruits competition. A 125-pound squash entered by C. Stein required two men to carry it into the exhibit hall. Congressman W. W. Bowers noted that there were no sheep in competition, even though they were in abundance in the county. Interest in the fair was manifested when San Diego held it in Balboa Park, and it was suggested that 100 acres be set aside for buildings and a racetrack. In subsequent years, Oceanside presented the fair. Media accounts had the fair being in Escondido from 1909 through 1912 then again in Balboa Park in the 1920s.

The opportunity for the fair to be located in a permanent location occurred in 1933, when Californians approved legislation that provided for pari-mutuel wagering on horse racing. The bill stipulated that revenue would go toward the sponsorship of state fairs. Sen. Ed Fletcher was receptive when James E. Franks and Frank G. Forward, active San Diego community leaders, journeyed to Sacramento and influenced Gov. James Rolph Jr. to approve the formation of a permanent county fair site. At that time, Fletcher owned the land where a golf course and riding stable operated close to rail lines and major roads, land that would later be bought for the fairgrounds.

The 22nd District Agricultural Association, a dormant entity, was reactivated in 1933 for the purpose of presenting the San Diego County Fair. Gov. Frank Merriam came to Del Mar and placed a cornerstone at the Mission Tower Plaza declaring the fair a major California Exposition. It became a construction project for the Works Progress Administration (WPA), created during the Depression to provide employment. While the federal agency provided the seed money, the agency had none for buildings and infrastructure.

William "Bill" Quigley, an investor from La Jolla, learned of the district's plight from James Franks and contacted Bing Crosby, who lived and kept several racehorses in Rancho Santa Fe. Crosby liked the idea of fronting $250,000 for building the grandstand and track. In partnership with Pat O'Brien, who lived in Del Mar, they formed the Del Mar Turf Club. They involved several Hollywood celebrities, including Joe E. Brown, Oliver Hardy, Gary Cooper, and others. Stock in the Del Mar Turf Club went for $100 per share. A 10-year lease for Thoroughbred racing was negotiated with the 22nd District. Since then, this has expanded to 20 years with renewal of the lease.

Sam and Joe Hamill, with Herbert Jackson, prominent San Diego architects, designed the grandstand and fair complex in a Mission Revival style. Meanwhile, the 22nd District was busy preparing for the inaugural fair. Franks became president, George Sawday was first vice president, Dr. N. Matzen second vice president, and Dan Noble became the secretary-treasurer. Directors included Frank Forward, Fred W. Mitchell, John Barger, and Robert Graham.

Fall dates of October 8 through 18 were selected for the fair's first run. It was the start of the rainy season, and for most of the 11 days it poured, flooding the grounds. Gov. Frank Merriam appeared to launch the fair, and a band concert by the U.S. Marine Corps at noon was followed by harness racing. Spanish fiesta girls performed in dazzling costumes, and Bunny Dryden, world-champion high-wire walker, went from the Midway to the grandstand on a single wire 110 feet above the ground without a net. There was a merry-go-round and Ferris wheel; a hoochie-koochie sideshow featured "Little Egypt," who promised to bare all inside a tent for a dime. She didn't really. A fireworks show in front of the grandstand ended the well-attended and sunny first day.

The fair was suspended during World War II. Crosby used the buildings for Bing's Bomber Builders, producing airplane parts. Army soldiers and marines camped in the stables when they were on overnight maneuvers. The first postwar fair was held in 1946. Paul T. Mannen, whose family owned Mannen Egg Company in La Mesa, was appointed to the board of directors and quickly assumed the leadership role. Ernest O. Hulick was hired as secretary-manager, and he brought in Fred Heitfeld from the California State Fair as special events director. Heitfeld came up with the Fairest of the Fair idea, and the first to hold that title was Gloria O'Rear of Del Mar.

Mannen, who had succeeded Hulick as manager, decided there should be a fair ambassador and that it should be someone who portrayed a Spanish don. The William Morris Theatrical Agency of Hollywood was providing some of the fair's entertainment, and it suggested Tommy Hernandez, who was appearing in some of the Cisco Kid films. He was given the role of Don Diego, a famous early California don whose name was Don Diego de Alvarado and in the early years lived on Tenth Street in Del Mar. In 1947, Don Diego and the Fairest of the Fair became the official greeters, visiting clubs and special functions throughout the Southland. They were featured in newspapers and magazines and appeared on radio and television.

Don Diego succumbed to leukemia in 1984, and the Fairest of the Fair program was discontinued in 2004. It was the longest-running queen pageant on record, with 58 contestants having been crowned, including Paramount starlet Marla English and actress Raquel Tejada Welch. During the late 1950s and 1960s, the fair was known as the Southern California Exposition as a means of attracting a wider audience. Later it was called the Del Mar Fair. It regained the title of San Diego County Fair in 1998 and now has an attendance in excess of 1.2 million and is considered a world-class event and destination.

This introduction was written by veteran newsman Bill Arballo, the third mayor of Del Mar and formerly the fairground's head of publicity. He is the last living witness to Governor Merriam's 1936 ceremonial laying of the fairground's cornerstone.

One

A FAIR IS BORN

When National City founder Frank Kimball was looking for an event to showcase the fledgling community's agricultural industry, he staged the first ever San Diego County Fair. Initiated and organized by the National Ranch Grange, a local association of resident farmers, the two-day event opened on September 22, 1880. So successful was the fair that Kimball later served as the state commissioner of horticulture. (Courtesy National City Library, Morgan History Room.)

THE FIRST ANNUAL

AGRICULTURAL FAIR HORTICULTURAL

FOR THE COUNTY OF SAN DIEGO,

Will be held at National City, Wednesday and Thursday, Sept. 22 and 23, 1880

UNDER THE AUSPICES OF NATIONAL RANCH GRANGE.

AMPLE preparations will be made by the erection of a suitable building, and nothing will be left undone that will in any way advance the interests of Agriculture, Horticulture and the Mechanical Arts in our County. The following Premiums are offered under and by authority of the Executive Committee, and it is earnestly hoped that every material interest of the County will be fully represented.

PREMIUM LIST.

ALL ARTICLES ENTERED TO BE PRODUCED OR MANUFACTURED BY THE EXHIBITOR.

Within three months, the National City fair had been planned and staged. School was closed, and the local paper wrote, "Everybody and his wife is going to the fair." Exhibits of fruit, vegetables, olives, embroidery, leatherwork, shell work, and paintings vied for premium prizes. "Amiable ladies . . . gave anyone a dinner such as they had never dreamed of for 25¢," reported the *National City Record* in its 30-column review. (Courtesy 22nd DAA.)

In 1889 and 1890, Escondido hosted the fall county fair on grounds located adjacent to and north of Washington Avenue; Charles A. McDougall was president. Former legendary gunslinger Wyatt Earp judged equestrian events, which included Silkwood, the big fast pacing stallion, and McKinney, the famous trotting horse, thrilling record crowds, many having arrived by special train following the 1888 completion of the Santa Fe Railroad branch line. (Courtesy Escondido Library, the Pioneer Room.)

Wyatt Earp

PANAMA-CALIFORNIA EXPOSITION 1915
San Diego California
View from Balboa Park

During the 1890s, the annual fall fairs were itinerant, sharing locations with Coronado, Oceanside, and San Diego. In 1915, the county fair was combined with the opening of the Panama California Exposition in Balboa Park, a popular venue that it would revisit in the 1920s and 1930s. (Courtesy 22nd DAA.)

More than anyone, nursery owner and city park planner Kate Sessions led the way for the beautification of San Diego through her horticultural expertise and in so doing championed the subsequent growth of the floriculture industry. By the 1930s, flower shows around the county had become an annual event and would be an important feature of the fair. In 1936, Sessions would be appointed to the committee for horticulture for the newly formed San Diego County Fair. (Courtesy 22nd DAA.)

In the 1920s, part of the San Dieguito lowlands was home to stables and a golf course. Owned by Ed Fletcher's South Coast Land Company, these 167 acres had been filled in to accommodate the Del Mar Golf Course, which ultimately failed because of saltwater seepage damaging the grass. Fletcher sold this land to the 22nd DAA for $25,000 to be used as a permanent site for the fair. (Courtesy 22nd DAA.)

The defunct golf club's pro shop got a new lease on life when the fairground complex was built. It was remodeled and became the entrance office to the early fairgrounds. During the war years, it housed interim manager Hazel Frasse and her family, and then for several decades it was used as part of the 22nd DAA's general offices. (Courtesy Bill Arballo.)

PREMIUM LIST

12TH SAN DIEGO
COUNTY
FAIR

BALBOA PARK
SAN DIEGO, CALIF.

October 7 to 11, Inclusive
1930

Open Tuesday, the 7th, at 6 P.M.

The 12th Annual County Fair was held in 1930, which suggests that there were several years when the fair event did not happen, since the inaugural fair was organized in 1880. The Great Depression following 1929's Wall Street crash was palpable to residents and businesses in San Diego County. However, in a show of optimism, the 1930 fair was staged. Exhibition categories included cattle, swine, sheep, goats, rabbits, poultry, pigeons, horticulture, vegetables, apiary, home goods, needlework, and 4-H exhibits. Premiums ranged from $10 for winners in livestock to $1 for winners in bedspreads. There would not be another fair until 1936. (Courtesy 22nd DAA.)

The first president of the 22nd DAA's board of directors was James E. Franks. Aided by Frank G. Forward of the Union Title and Trust Company and later by purchasing Ed Fletcher's land, Franks would get the approval from then-Governor Rolph to organize a permanent county fairground. This 1930s photograph captures Franks in front of his Mission Beach drugstore, Franks Sundries, which was managed by his wife, Gladys. (Courtesy Jimmie Marie English.)

James E. Franks will go down in history as the true father of the fair. Without his leadership, the permanent Del Mar location would not have become a reality. In this late-1930s family photograph taken in Mission Beach, Jim Franks, as he preferred to be known, stands with his three daughters, (from left to right) Lucille, Doris, Jimmie Marie, and his wife, Gladys (far right). A son, Gerald, is not shown. (Courtesy Jimmie Marie English.)

14

Sen. "Colonel" Ed Fletcher, who was a force in the development of San Diego County, had worked with Jim Franks in the subdivision of the Lake Cuyamaca area. Fletcher's brother-in-law Eugene Batchelder, a Del Mar resident, worked as an agent for Fletcher's South Coast Land Company. It was through his agency that the 22nd DAA was able to purchase the land for the fair site. Fletcher was appointed as a member of the fair's reception committee for its opening year. (Courtesy Jill Coughlin.)

By 1936, seven permanent members were appointed to the board of directors for the 22nd DAA: James E. Franks, president and general manager; Frank G. Forward, vice president; George Sawday, director; Robert Graham, director; Dr. N. Matzen, director; Fred Mitchell, director; and John Bargar, director. D. A. Noble was appointed as secretary/treasurer. Eugene Batchelder was an original appointee but was replaced by Matzen because of a conflict of interest. (Courtesy 22nd DAA.)

15

The Hamill brothers, Sam (left) and Joe, inspect the plans for the building project, an undertaking financed by the Works Project Administration. With civil engineer Herbert Louis Jackson and fellow architect Albert Owen Treganza, the team was the inspiration behind the Mission Revival–style architecture. Both Treganza and Hamill had worked under the tutelage of architect Lilian J. Rice, resident architect for Rancho Santa Fe. (Courtesy Del Mar Turf Club.)

This preliminary master-plan graphite sketch shows Sam Hamill's City Beautiful–inspired layout with its central thoroughfare and buildings leading from it. The Mission Revival–themed architectural style was chosen as a symbol of the county's colonial Hispanic roots. Although ultimately not all of these buildings were included because of funding shortages, the integrity of the overall design was maintained. (Courtesy 22nd DAA.)

In 1936, Bing Crosby, internationally renowned entertainer, had a preliminary meeting with Jim Franks at the Huntington Hotel in Pasadena to discuss organizing horseracing at the fairgrounds. A second meeting at the Beverly Wilshire ultimately led to the formation of the Del Mar Turf Club. Crosby was captured in this 1937 shot during the first year of Thoroughbred racing held at the newly opened racetrack. (Courtesy International Club Crosby.)

Fred Mitchell, second from the left in this 1930s Rotarian group shot, owned the *Carlsbad Journal* and as such was appointed as the fair's chairman of publicity. A director, he also served as chairman for the Premium List Committee and a member of the Concessions Committee. As director of the Floriculture Committee, he was paid $1,600, a fact that would later cause editorial attacks by the *San Diego Sun*. (Courtesy Carlsbad City Library, Carlsbad History Room.)

All the hard work paid off for Jim Franks and his directors when, on August 20, 1936, Governor Merriam—looked on by Bing Crosby and aided by state WPA administrator Frank McLaughlin—ceremoniously laid the cornerstone of the fairgrounds' half-a-million-dollar project. They were at the site of the Bell Tower, or El Campaniaro. Work crews had worked furiously to construct the fair by the October 1936 opening. (Courtesy 22nd DAA.)

October 8, 1936, was opening day. It began with Merriam arriving dramatically by stagecoach. Entertainment, exhibits, rides, and contests thrilled 50,000 attendees. The Big Wheel dominated the Midway, where Bunny Dryden walked a 110-foot high wire. Tents housed sideshows, stunt artists, food vendors, and amateur contests. The first fair beauty queen was selected from 10 finalists and crowned on the fair's last Saturday night. (Courtesy 22nd DAA.)

Other attractions included daredevil motorcycle riders, vaudeville and circus acts, children's programs, and the famed Slide for Life, "a daring, thrilling performance from the summit of the grandstand to the midway with a human hurtling through space on a wire," according to the fair's program. There were two music shows daily, which included concerts by the U.S. Marine Band, the National Guard Band, the Naval Training Station Band, and the Bonham Brothers Boys Band. (Courtesy 22nd DAA.)

This 1936 postcard shows the beauty of the Plaza de Mexico. Its design recaptured the "spirit of the viceroys, the rancheros and landowners and the gifted Franciscans," Sam Hamill was quoted as saying. Design modifications required by the WPA called for simpler forms that reflected a mission style. "We went from the more elaborate forms of the Spanish buildings to the simpler church forms of the Fathers." (Courtesy 22nd DAA.)

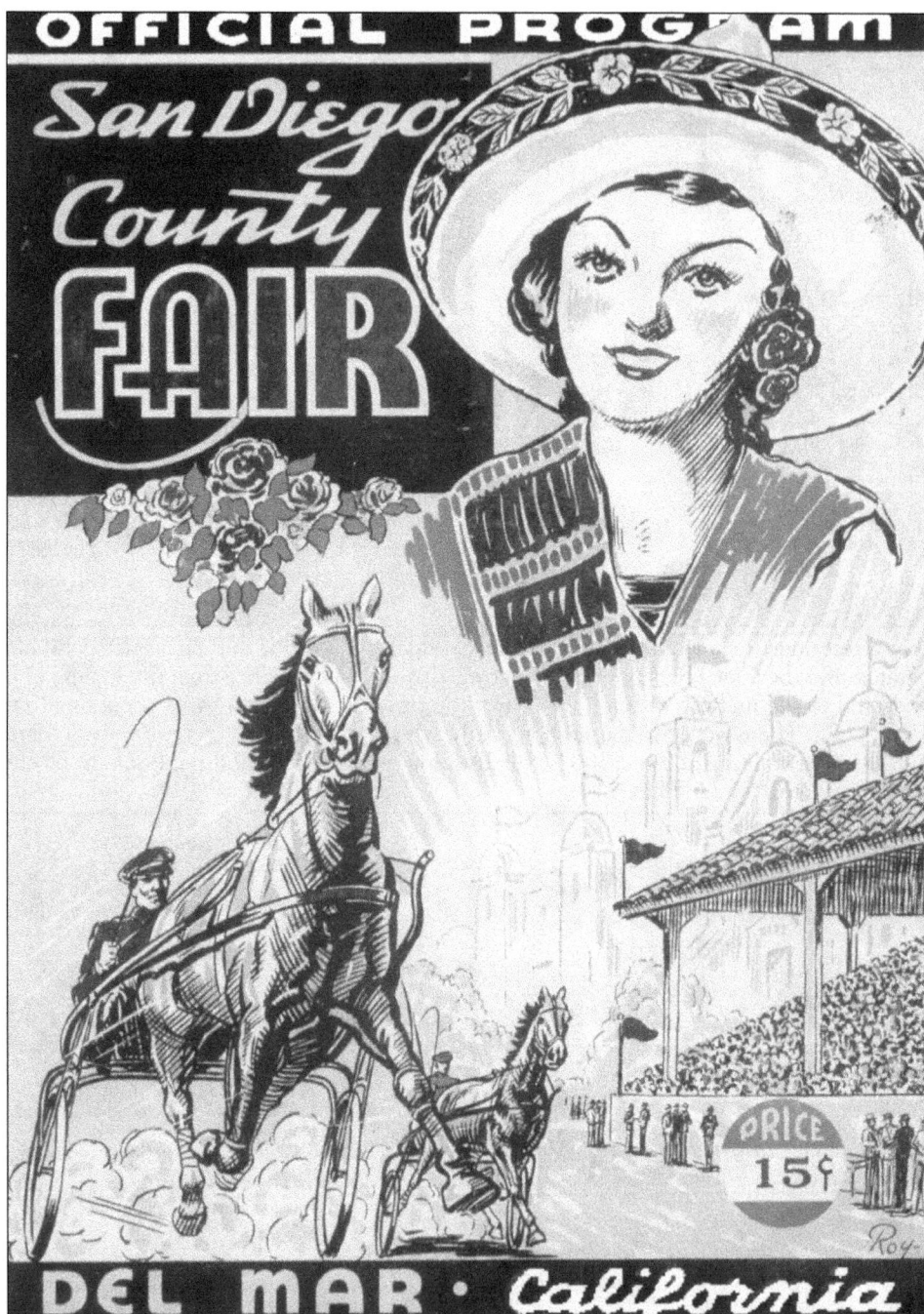

The directors voted to organize the second year's fair as a summer event. In 1937, it ran from August 7 through 15 in an effort to avoid the rains that had plagued the prior season. This nine-day run also facilitated the earlier dates set for the inaugural racing season. This official program showed harness racing—staged on the newly completed mile-long track—and spotlighted the Mexican flavor of the fair's hospitality by picturing a welcoming senorita. A marketing campaign to publicize the summer fair utilized California's historic romantic past as a colorful element in creating a unique image for the fairgrounds. (Courtesy 22nd DAA.)

20

When Sam Hamill chose the distinctive Mission Revival style of architecture, he was echoing a trend to not only uphold the Spanish colonial roots of California but also to add integrity to a style that was considered regional. The arches, parapets, bell towers, and restrained ornamentation that typify this style also added a romantic element to the entire fairgrounds complex that remains today. (Courtesy 22nd DAA.)

The Mission Bell Tower building soon became a focal point of the fairgrounds complex, with people meeting at its base to sit and enjoy the tropical landscaping that was an important feature of the overall architectural plan. It was also used as one of the exhibition halls; in the recent evacuations during the firestorm of 2007, it was used as a temporary hospital for evacuated patients. (Courtesy 22nd DAA.)

This 1936 aerial view shows the scope of the fairgrounds complex, which became a most costly project. An average of 380 men worked on construction daily with hundreds of thousands of adobe bricks being manufactured on site. An adobe wall was built around the entire ground's perimeter. In 1937, the WPA abandoned the project, leaving $200,000 in work yet to be done; Bing Crosby and Pat O'Brien funded this shortfall. (Courtesy 22nd DAA.)

On the night of Sunday, October 18, the gates closed on the last of the happy but tired fairgoers. The parking lots had been packed, and attendance had surpassed the directors' expectations. There had been several days of torrential rains during the 11-day run when unpaved grounds had turned to a sea of mud. But despite this, attendance was unprecedented, and the press hailed the first fair as an unqualified success. (Courtesy 22nd DAA.)

Two

DON DIEGO AND THE FAIREST OF THE FAIR

The crowning of the Fairest of the Fair by Don Diego was the high point of the entertainment offered during the county fair's summer run. Raquel Tejada, now renowned celebrity Raquel Welch, has the distinction of being an internationally recognized star. Her fairgrounds publicity photographs are the most often requested of all the 22nd DAA's archived collections. The year 2008 celebrates 50 years since Welch owned her title of Fairest of the Fair. (Courtesy 22nd DAA.)

Barbara Watson made San Diego County fair history when she became the first ever beauty queen crowned at the fair in 1936. A San Diego resident, she competed with girls from across the county who vied for the title. Rules dictated that they had to be between 18 and 25 years old and a county resident for at least six months prior to the competition. The title was then known as Queen of the Fair. (Courtesy 22nd DAA.)

Gloria O'Rear
Del Mar
1946

Following an absence of fair beauty contests during the late 1930s and throughout the duration of World War II, Gloria O'Rear made fair history when in 1946 she was crowned the first Fairest of the Fair, a title created by the fair's newly formed publicity and marketing department. O'Rear, 18, was an Oceanside Junior College student. (Courtesy 22nd DAA.)

By 1947, it was decided that the Fairest of the Fair would be a regular annual event with several month's run up of qualifying rounds to determine lucky finalists. The media helped to publicize contestant sign ups by printing entry forms in their papers. Claire Le Roy of East San Diego was ultimately declared winner at the final round on the last Saturday night of the 1947 fair season. This was also the first year that Don Diego crowned the winner. (Courtesy 22nd DAA.)

In 1948, Oceanside's Jeanene Cross, 17, was crowned the Fairest of the Fair after competing against nine contestants in her qualifying round as the "fairest equestrienne." A petite 5-foot-2-inch-tall brunette with blue eyes and 105 pounds in weight, she had previously reigned a year earlier as the Fiesta Princess in the Days of San Luis Ray celebration in Oceanside. (Courtesy 22nd DAA.)

Alice Kerr, a graduate of Oceanside-Carlsbad High School, became Fairest of the Fair in 1949. She was chosen from girls who represented every school district in the county. To be eligible for entry, contestants were required to have a B average in their subjects and possess personality, charm, and beauty. Alice had a B-plus grade point average for her four years of high school and presumably all the other requirements. (Courtesy 22nd DAA.)

Pat Beasley of Linda Vista was the fair queen who ushered in the new decade of the fair's famed beauty contest when she was crowned in 1950. Contestants then were judged on character, poise, personality, intelligence, charm, and beauty of face and figure. By now, the competition was widely recognized, and contestants became celebrities in their own right—at least on a local basis. (Courtesy 22nd DAA.)

North Park's Marla English, competing as a stenographer, was crowned in 1951. She was not only the youngest ever, at 16, to become Fairest of the Fair but was the first to take her title beyond the local reach of San Diego County. Her professionalism, congenial personality, and jaw-dropping beauty caught the attention of Hollywood, who signed her to a major movie contract following her reign at the fair. (Courtesy 22nd DAA.)

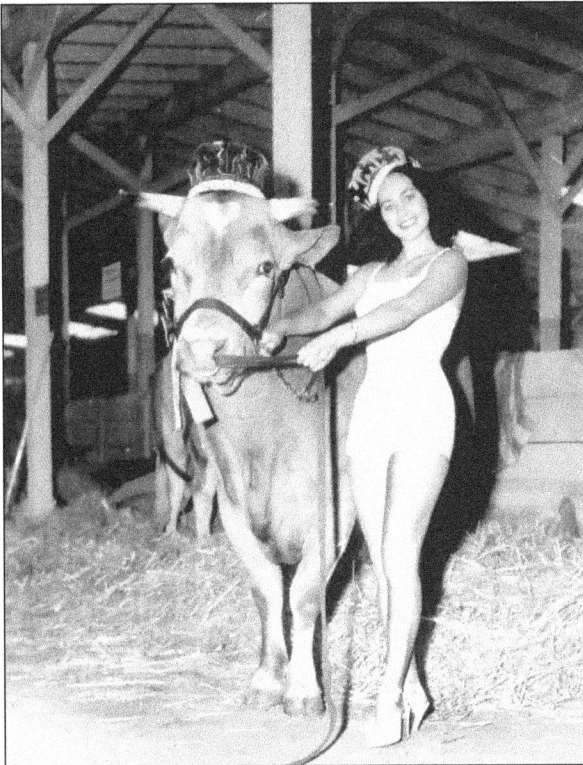

Marla would soon be appearing with movie stars rather than mooing stars when she was contracted with Paramount to appear with Spencer Tracy in a film being made in Switzerland. Some say her mother blocked Marla from making the film, while others cite an infection caused by inoculations, which were necessary for the overseas visit. Either way, Marla was unable to star in the film, and her Paramount contract was duly canceled. (Courtesy 22nd DAA.)

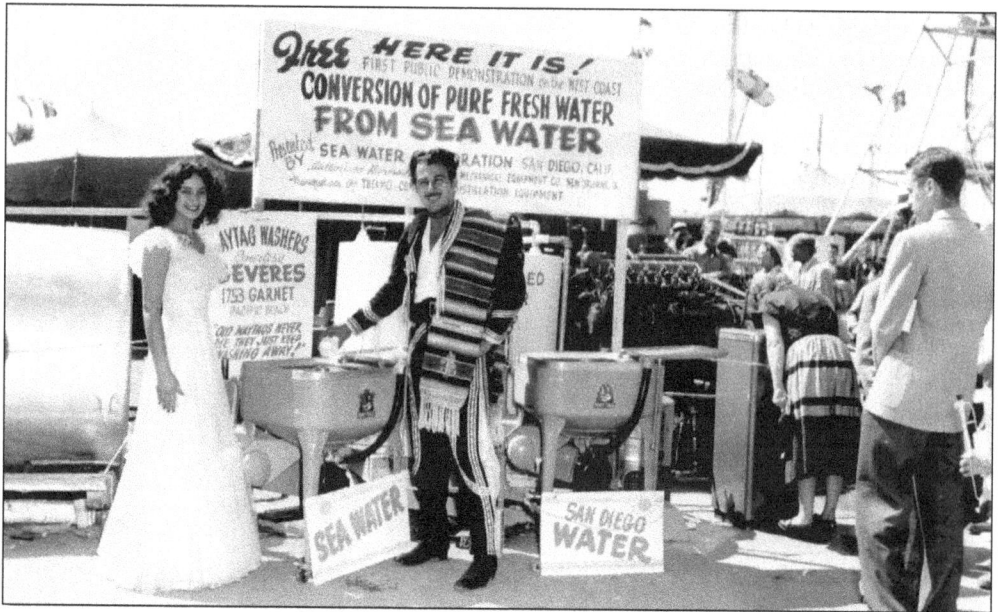

During the queen's reign, daily tours of the fairgrounds' many exhibits offered more opportunities for promotional shoots. Marla and Don Diego pose by San Diego's Sea Water Evaporation Company's exhibit. Utilizing old Maytag washing machines, they demonstrated the turning of salt water into fresh drinking water, which was then hooked up to a drinking fountain for willing onlookers to sample. (Courtesy 22nd DAA.)

As well as publicity shots with livestock and exhibitors, the fair queen and Don Diego appeared with celebrities who had been booked as part of the lineup of the fair's grandstand entertainment. Joining them in 1951 were famed guitarist Les Paul, known world wide for his superior-quality electric guitar manufacturing, and his wife and stage partner, vocalist Mary Ford. (Courtesy 22nd DAA.)

In 1952, Mae Entwistle of San Diego was crowned Fairest of the Fair. The evening gown had been replaced by the bathing suit as the suitable attire for the crowning ceremony. Flowers, a trophy, and promotional gifts were added to the prizes offered to the winning contestant and her runners up, who were known as "her court of lovelies." (Courtesy 22nd DAA.)

Part of the fair queen's crowning ceremony was the symbolic passing of the title by the former queens. This group shot includes Marla English to the right, Mae Entwistle to the left, and at center 1953's newly crowned Dorothy Wanket, noted as representing San Dieguito, a region that encompasses Del Mar to the south, Encinitas to the north, and Rancho Santa Fe to the east. (Courtesy 22nd DAA.)

The romantic notion of a dashing caballero playing the welcoming host to the county fair was solidified by the 1950s. The colorful character of Don Diego, posing here in 1953 with Dorothy Wanket, became an enduring symbol of the hospitality of the fairgrounds and of the 22nd District Agricultural Association and was used widely in publicizing the upcoming fair. (Courtesy 22nd DAA.)

A high-heeled, bathing-suited beauty photographed with a neatly shorn sheep may seem mismatched today, but for 1950s publicity shoots it was neither novel nor incongruent. The origins of the county fair as a venue to showcase and judge not only the agricultural fruits of the county but also the livestock of the area farmers made the pairing of Dorothy Wanket with a sheep natural and appealing. (Courtesy 22nd DAA.)

In 1954, Jo Johnson of Renaldo Village was crowned Fairest of the Fair. Competing against 21 hopefuls, she was crowned at the nearby Hotel Del Mar, winning $100 in cash plus valuable merchandise prizes. She was the first queen to have the dual title of Fairest of the Fair and Miss Con-Tour, the name given as the San Diego Convention and Tourist Bureau's official hostess. (Courtesy 22nd DAA.)

According to local papers, for the first time in the 10-year pageant history a girl from Point Loma was crowned Fairest of the Fair in 1955. Gay Cowie, 17, "a blonde haired, blue eyed beauty" had performed with her sisters as the Gay-Bon-Lin Trio, having recorded songs for the Combo label before winning the title. She had won a preliminary event sponsored by the Ocean Beach 20/30 Club to qualify for entry into the pageant. (Courtesy 22nd DAA.)

Publicity shoots often included the Fairest of the Fair, in this case Gay Cowie, being driven around the fairgrounds complex. In this 1950s photograph, she is joined by Don Diego (left) and Lawrence Welk (right). Welk, born in 1903, was a musician, accordionist, bandleader, and television impresario, hosting *The Lawrence Welk Show* from 1951 to 1982. His style came to be known to his large number of fans as "champagne music." (Courtesy 22nd DAA.)

Only a year after her 1956 crowning as the Fairest of the Fair, National City's Patricia Lou Thompson was married. She became the wife of Carr Beebe Jr. of San Diego. She was the daughter of Rev. George S. Thompson, who officiated over her wedding at the First Baptist Church in Chula Vista. (Courtesy 22nd DAA.)

In 1957, Pamela Britton and Arthur Lake, costars in the television show *Blondie and Dagwood*, among others, judged Ellen Marie Emig to be the 18th Fairest of the Fair and Miss Con-Tour. As winner, she received a "three day holiday in Las Vegas, $300 in cash, season credentials to the Del Mar Races, a swimsuit of her choice, and all expenses paid during her reign at the county fair," according to a *San Dieguito Citizen* newspaper article written during the 1957 summer fair. (Courtesy 22nd DAA.)

Raquel Tejada had beauty, brains, and talent. She took dancing lessons as a child and was winning beauty pageants by the time she was a teenager. Among her titles were Miss Photogenic, Miss La Jolla, and Miss San Diego. An honor roll student, after graduating from La Jolla High School she entered college on a theater arts scholarship. In January 1959, she secretly married fellow high school student James Welch. (Courtesy 22nd DAA.)

After the 1958 fair season, Raquel took on more professional roles. In 1959, she played the lead in the *Ramona Pageant*, a yearly outdoor play at Hemet, California, based on Helen Hunt Jackson's novel *Ramona*. She then became a local television weather forecaster, a model for Neiman Marcus, and in 1966 starred as Loana in the movie *One Million Years B.C.*, which led to her status of superstar in the 1960s and 1970s. (Courtesy 22nd DAA.)

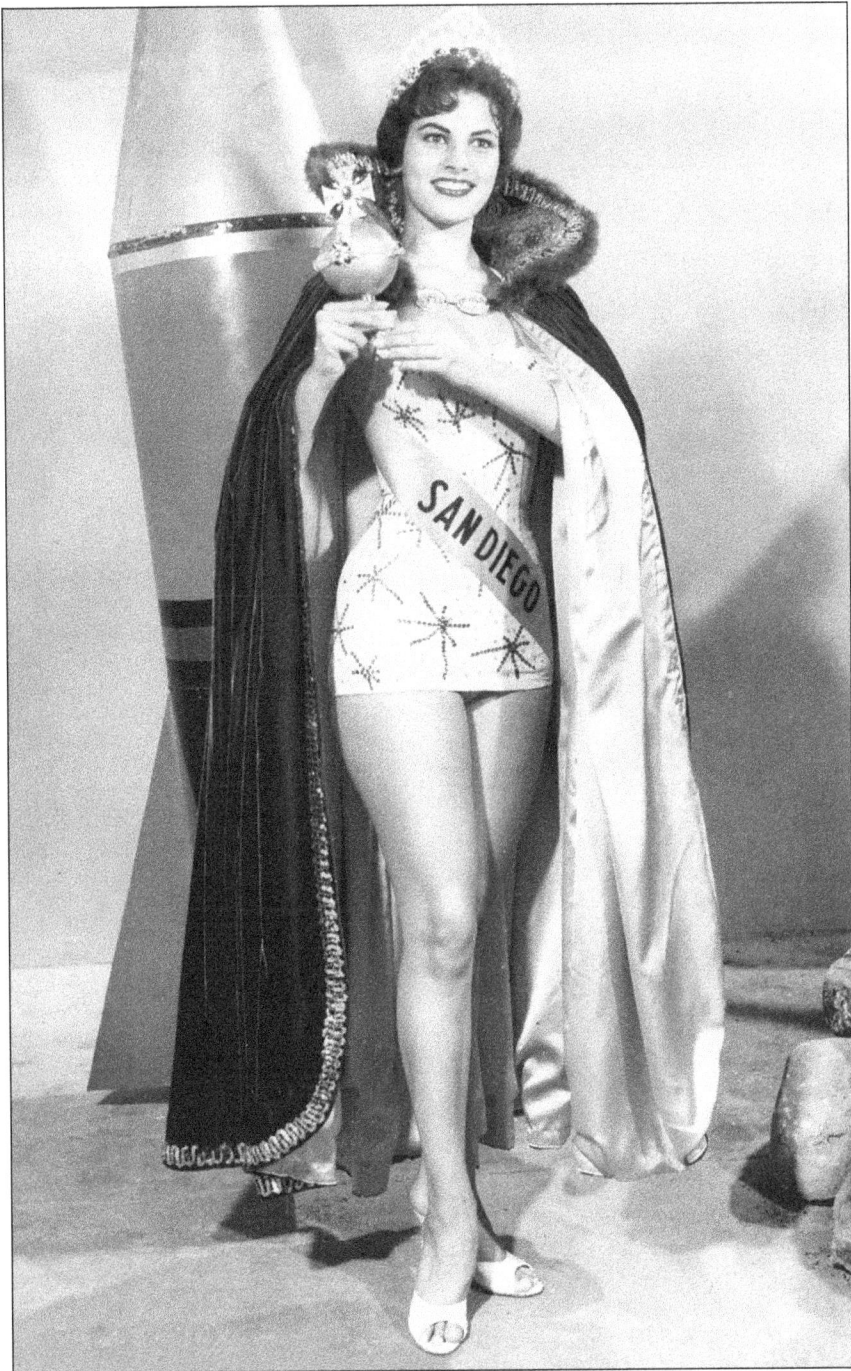

Raquel Tejada was crowned Fairest of the Fair and Miss Con-Tour in 1958. Born in Chicago, Illinois, in 1940, she was the oldest of three children born to Armando Carlos Tejada Urquizo and Josephine Sarah Hall. Her father, who emigrated from Bolivia, was an aerospace engineer of Spanish-Castilian descent; her mother was an Irish American. When she was two years old, her father was transferred to San Diego, California, and the family moved to La Jolla. (Courtesy 22nd DAA.)

In 1959, Shirlee Brown of Ramona would be the final 1950s representative of that decade's Fairest of the Fair. The successful run of the much anticipated pageant would continue through the next four decades but then would stall as the new millennium got under way. By 2004, the pageant was relegated into the past. (Courtesy 22nd DAA.)

As part of the variety entertainment that appeared on the grandstand stage, audiences were thrilled by circus-style acts and acrobatic dancing troupes. This 1950s beaded and bejeweled bikini-clad team of four women integrated a Native American theme into their dance routine. (Courtesy 22nd DAA.)

"Ice Lady" Karyl Leigh was featured in the Icelandia spectacular musical skating review, which opened at the fairgrounds on June 24, 1957. The show, presented on a portable rink, ran for 11 matinee and evening performances during the fair. Leigh was considered one of "the headliners in the tuneful, gag-filled ice show," according to the *San Dieguito Citizen*. Note that her name is misspelled in the photograph. (Courtesy 22nd DAA.)

Don Diego is shown with the famous "Wazzans from Tangiers," who were billed as the world's greatest acrobatic team and a major attraction of the Polack Brothers International Circus, owned by I. J. Polack and partner Louis P. Stern. The Polack show was unique in that it attracted world-class acts and was able to exhibit in large arenas or outdoor venues. (Courtesy 22nd DAA.)

Paul Albert Anka (right), born in 1941, was a Canadian singer, songwriter, and actor of Lebanese origin. His first hit, "I Confess," was launched when he was 14 years old. In 1957, he went to New York auditioning for Don Costa at ABC, singing "Diana," which brought Anka instant stardom. Reaching No. 1 on the charts, it is still considered the best-selling 45 in music history. Anka appeared at the fairgrounds in 1965. (Courtesy 22nd DAA.)

Charlie McCarthy (center) will forever be remembered as the best-known dummy on radio. As a teenager in the late 1910s, ventriloquist Edgar Bergen (right) purchased the wooden dummy, and they remained partners until Bergen's death in 1978. Edgar and Charlie were given their own show, which was broadcast on the air until 1956. Their success was a result of the hilarious and somewhat risque conversations that Charlie had with stars like actress Mae West and comedian W. C. Fields. (Courtesy 22nd DAA.)

The dance troupe Danza Azteca de Anahuac, a team of three men and two women from northern New Mexico, helped preserve the ancient cultural traditions of Mexico through their act, which they staged at the 1959 fair. The dancers described the meaning of the rituals and beliefs they depicted in performance and talked about the influence of this cultural history on their daily lives. (Courtesy 22nd DAA.)

Andy Griffith, seated to the right on the diving board, is one of television's most personable and enduring stars. He is perhaps best known as Andy Taylor, the central character in *The Andy Griffith Show*, which aired on CBS from 1960 to 1968 and consistently ranked among the top-10 shows in each of its eight seasons. As a down-home attorney in the even longer running *Matlock*, Griffith continued to make a unique contribution to television Americana. (Courtesy 22nd DAA.)

Singer-songwriter Ricky Nelson (with guitar) appeared at the fairgrounds in 1967. He was one of the most popular teen idols of the late 1950s and early 1960s. His hits included "I'm Walkin'," "Travelin' Man," "Fools Rush In," "It's Late," and "Garden Party." Most of these songs were originally introduced on *The Adventures of Ozzie and Harriet* television show, in which he starred. He tragically died in a private plane crash on New Year's Eve 1985. (Courtesy 22nd DAA.)

Local celebrity and world-known movie star Leo Carillo (right) was a regular feature at the fair in the 1940s. This shot shows him with Don Diego performing on stage during the acclaimed Fiestacade, a variety show extravaganza that entertained fairgoers twice a day with matinees and evening shows. (Courtesy 22nd DAA.)

The fictional character Don Diego lives on in the minds and hearts of those who remember him. His greeting, "Bienvenidos," symbolized both the cultural heritage of the fair and its friendly welcome. This clock tower, built in the 1950s, is embellished with a tiled mosaic of Don Diego; a statue of the handsome caballero also greets attendees at the fairgrounds entrance—both immortalizing his spirit. (Courtesy 22nd DAA.)

Over the almost four decades that Tommy Hernandez, in his role as Don Diego, welcomed patrons to the fair, he gained celebrity status in his own right and had a huge following. This 1979 certificate from Man Watchers, Inc., is in recognition of his contributions to the entertainment industry and proclaims Don Diego a man "well worth watching." (Courtesy 22nd DAA.)

Man Watchers, Inc.

THE "WELL WORTH WATCHING" AWARD

presented to

Don Diego

on the 21st day of June, 1979

for being chosen

San Diego's most Watchable Host

Suzy Mallery
President,
Man Watchers, Inc.

"It's our turn-ON, now"

A 1950s Grandstand star was Ina Ray Hutton, a nationally known celebrity famed as an American female bandleader during the big band era. A Chicago native born in 1916, she began dancing and singing in stage revues at the age of eight, a trend she continued though the 1930s. In 1934, she was asked to lead an all-girl orchestra, which was disbanded five years later; in 1940, she led an all-male orchestra. (Courtesy 22nd DAA.)

Canadian-born actress and singer Gizelle (often spelled Gisele or Giselle) MacKenzie had a career in entertainment that spanned 45 years. Getting her break on *Hit Parade* as a pop star, she then went on to star in comedy shows, variety shows, quiz shows, and movies. She performed at the fair as a vocalist in 1965. (Courtesy 22nd DAA.)

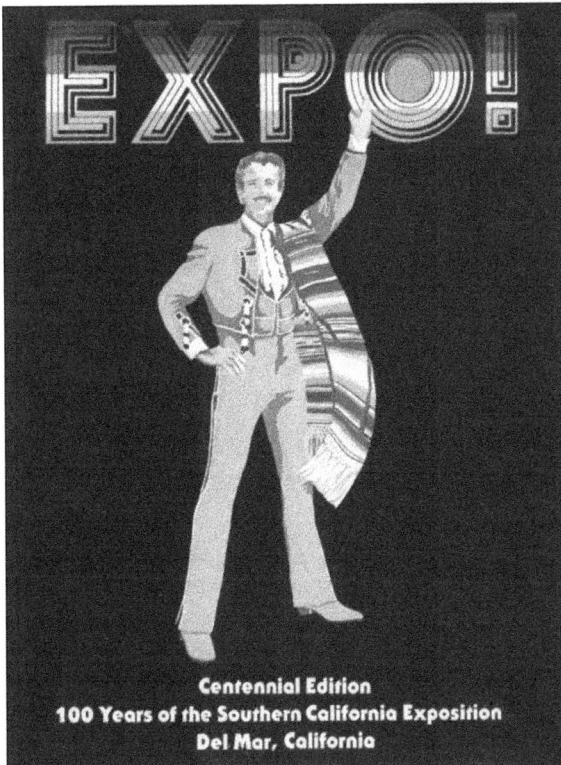

EXPO!

Centennial Edition
100 Years of the Southern California Exposition
Del Mar, California

Tommy Hernandez's role as Don Diego spanned an amazing 37 years, a run that is considered the longest continual gig in entertainment history. His character graced the cover of the celebratory program for the 100th anniversary of the county fair in 1980. Known then as the Southern California Exposition, the milestone gave a sense of tremendous county pride. Hernandez sadly died of leukemia just four years later. (Courtesy 22nd DAA.)

Kari Stout
1994 Fairest of the Fair
Miss San Diego County/
Miss Santee

Photo by Light Impressions

The Del Mar Fair
Presents

50th

ANNIVERSARY
1995
Fairest of the Fair
Pageant

The year 1995 was celebrated as the 50-year anniversary of the Fairest of the Fair pageant. Thousands of area hopefuls had competed during its half-century run. Many of the contestants went on to garner careers in the entertainment field. However, by 2003, the pageant had reached the end of the runway, and beauty contests were considered outdated. The final Fairest of the Fair was Carlsbad's Troy Lemperle, who held the title for the final year in 2004. (Courtesy 22nd DAA.)

Three

THE DEL MAR TURF CLUB

Entertainer Bing Crosby was the guiding force behind organizing and partially funding the Del Mar Turf Club. On the racetrack's opening day, he broadcast a 40-minute radio show for NBC. It featured an account of the running of the Inaugural Handicap Race, his signature track song "Where the Surf Meets the Turf," and celebrity guest interviews, a practice that would continue through the 1937 season. (Courtesy International Club Crosby.)

Landscaping remains unfinished in this 1937 shot showing the rear of the grandstand. Bing Crosby is captured with architect Joe Hamill examining a watercolor rendering of the buildings. It was a frantic scramble to complete construction by July 3. Media sources claimed that the paint was still wet on several of the buildings when opening day arrived, and that construction on other fairgrounds buildings continued around the crowds. (Courtesy International Club Crosby.)

Crosby's love of Thoroughbreds brought him in partnership with Lindsay Howard. Together they formed Binglin Stables, which Crosby located on his Rancho Santa Fe estate. Designed in 1934 by architect Lilian J. Rice on the site of the old Osuna adobe hacienda, it became 97 acres of a state-of-the-art breeding ranch. Captured in this classic shot is a relaxed Crosby with a Binglin Thoroughbred. (Courtesy International Club Crosby.)

Opening Day for the Del Mar Race Track was on Saturday, July 3, 1937. The first patron through the fairgrounds gates at 10:20 a.m. was San Diego's Mrs. W. R. Robinson, greeted by a relaxed, casual, pipe-smoking Crosby. He surprised attendees by playing ticket collector and was looked on by a sizable crowd of well-placed media personnel. (Courtesy Del Mar Thoroughbred Club.)

Captured in motion on opening day are Crosby (center), O'Brien (right), and track manager Bill Quigley (left), who mingled with customers touring the racetrack complex prior to the inaugural race. At 2:00 p.m., formal opening ceremonies were underway. O'Brien, flanked by two marines, raised the American flag as Crosby led the band, which played "Stars and Stripes Forever." (Courtesy Del Mar Thoroughbred Club.)

The steward's stand doubled up as a bandstand on opening day at the racetrack. Prior to the event, a complete dress rehearsal was staged to assure a problem-free day. Not only did the band rehearse, but cashiers at the pari-mutuel windows also practiced, receiving bets, familiarizing themselves with the totalizator—a wager ticket vending system—and operating state-of-the-art electric cash registers. (Courtesy Del Mar Thoroughbred Club.)

This view of the grandstand shows wall-to-wall spectators who gathered anxiously waiting for the first race, which was scheduled for 2:15 p.m. but actually went off at 2:24 p.m. They wagered a total of $183,015 that day on the eight-race program. In addition to betting, crowds were also able to pick out scores of movie stars who, in a show of support, attended the racetrack that day. (Courtesy Del Mar Thoroughbred Club.)

These horses are being led alongside the grandstand building toward the track on opening day, illustrating a conscious design decision by Sam Hamill to keep the horses in full view during this trek. The elegant beauty and serenity of the horses and their jockeys, juxtaposed with the beautiful Mission architecture, demonstrate Hamill's creative expertise. (Courtesy 22nd DAA.)

The first horse to cross the finish line in the racetrack's inaugural race, sporting the Binglin blue and gold colors, was Crosby's horse High Strike. Ridden by jockey Jackie Burill, the two-year-old gelding caused spectators to go wild as it claimed its place in history, leading all the way to win. (Courtesy Del Mar Thoroughbred Club.)

One of the decisive factors to the fairgrounds' location in Del Mar was its proximity to major roads and the railroad. The Atchison, Topeka, and Santa Fe Railroad constructed a spur track that ran directly to the fairgrounds to transport patrons and horses to the track. For a $2 round-trip ticket, passengers were able to arrive trackside from Union Station in Los Angeles in time for the first race. (Courtesy Del Mar Thoroughbred Club.)

The paddock area drew a crowd of onlookers who could observe the grace and power of the Thoroughbreds as each one was paraded for display, a tradition that continues today. Olive trees from nearby Olivenhain, just north of Rancho Santa Fe, added a little shade and landscape interest to the area. (Courtesy Del Mar Thoroughbred Club.)

50

The press box was a hive of activity during the racing meets. In this shot, Bing Crosby, shown at the typewriter with his signature pipe, is flanked by the San Diego sports editor George Herrick. Turfwriter Maurice Bernard is in the foreground, and *San Diego Union* writer Harry Hache is standing behind them. (Courtesy Del Mar Thoroughbred Club.)

The business of horseracing and entertainment was a family affair for the Crosbys. Photographed trackside at a Turf Club table are Crosby, top right, with his father, Harry L. Crosby Sr., to his left, and his mother, Kate, is shown in the bottom right next to his wife, Dixie Lee, bottom left. Crosby Sr. helped manage Bing Crosby's business and estate and lived at his Rancho Santa Fe estate. (Courtesy Del Mar Thoroughbred Club.)

Celebrities were often called upon and were willing to present trophies to winning jockeys. It kept them in the public's hearts and was an opportunity to ensure media attention. Lucille Ball, second on the left, and her husband, Desi Arnaz (far right), are posing with the winning jockey and the winning horses owner and wife during the 1940s season. (Courtesy Del Mar Thoroughbred Club.)

Crosby's celebrity friends included actor and comedian W. C. Fields, captured here examining the day's card. Born William Claude Dukenfield in 1880, Fields was known for his on-screen persona who "teetered on the edge of buffoonery but never quite fell in; an egotist blind to his own failings; a charming drunk; and a man who hated children, dogs, and women—unless they were the wrong sort of women," as noted in a commentary by Ronald J. Fields in 1949. (Courtesy Del Mar Thoroughbred Club.)

Winning jockey Owen Webster is congratulated by movie actor Edward G. Robinson and 1940s movie starlet Joan Bennett, who presents him with the customary floral bouquet. Webster had just won the 1937 Coronado Handicap, riding Boss Martin to victory. (Courtesy Del Mar Thoroughbred Club.)

Not everyone left the day's races a winner. Comedian Red Skelton gives a literal demonstration of "losing his shirt" after a day at the track. Skelton, born in Indiana in 1913, began his entertainment career as a circus star at age 15. A successful four-decade-long career followed, with Skelton performing on radio, television, and in the movies. In 1986, he earned the Academy of Television Arts and Sciences' Governor's Award. (Courtesy Del Mar Thoroughbred Club.)

There was a major marketing advantage to having a world-known celebrity hold the reins of the Del Mar Turf Club. It meant that several high-profile entertainers support his endeavor. Part of the track's attraction was star spotting. Lucille Ball and Desi Arnaz, America's funniest and most celebrated husband-and-wife team during the fledgling years of television, were a popular couple at the track. (Courtesy Del Mar Thoroughbred Club.)

In 1938, actress Ann Sheridan, famed Warner Brothers' starlet, who like Betty Grable was a popular 1940s pin-up girl, presented the winning trophy to trainer Bill Finnegan after the La Jolla Handicap was won by Dogaway that year. Looking on are actor Eddie Norris and jockey Eddie Yager. (Courtesy Del Mar Thoroughbred Club.)

The racetrack also attracted celebrity sport stars like famed heavyweight champion Jack Dempsey, pictured during a track visit in 1937. From left, posing with Dempsey, are C. Pierce, Armando Fermin, George Burns, Owen Webster, and Bobby Varner. (Courtesy Del Mar Thoroughbred Club.)

Jimmy Durante was immortalized when the City of Del Mar proclaimed him honorary mayor and named the road to the fairgrounds in his honor. The street sign for Jimmy Durante Boulevard is flanked from left to right by then–Del Mar Fairgrounds manager Bob Jones, Del Mar mayor Vic Koss, Jimmy Durante, and Del Mar deputy mayor Bill Arballo, who also served as head of the fairgrounds' publicity relations. (Courtesy Bill Arballo.)

Preparing for the legendary race the morning of Friday, August 12, 1938, George Woolf puts Seabiscuit through his paces for one of the most historic horse races ever run: Seabiscuit *v.* Ligorati. The two-horse race did not meet minimum entry requirements, and therefore, special approval was granted by the California Horse Racing Board. The approval provided that there was to be no wagering, that the public must be informed concerning any possible cancellation of the contest, and if the horses did not run, patrons should receive admission refunds. (Courtesy Del Mar Thoroughbred Club.)

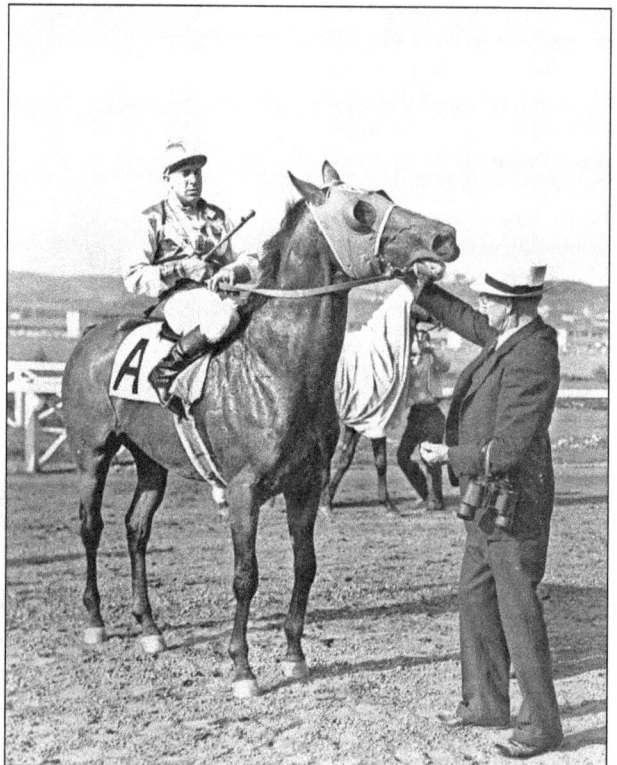

The famed Binglin's Ligaroti race against Charles Howard's Seabiscuit would serve to put the Del Mar Race Track officially on the map. With a $25,000 winner-take-all purse, this two-horse race was watched by a record crowd of 20,000 and nationwide radio audience. Here trainer Tom Smith greets Seabiscuit, ridden by renowned jockey George "The Iceman" Woolf, who won by a nose. (Courtesy Del Mar Thoroughbred Club.)

This action shot captures the head-to-head challenge of the six-year-old Argentine import Ligaroti with the "rugged old Seabiscuit, the reigning handicap champion of the West" ahead by a nose, as noted in *Del Mar, Its Life and Good Times*. Bill Quigley, general manager, proposed the race, which also pitted father Charles Howard, owner of Seabiscuit, against his son Lindsay Howard, co-owner of Ligaroti. (Courtesy Del Mar Thoroughbred Club.)

After the historic race, Dixie Lee Crosby presents the winning trophy to Charles S. Howard, Seabiscuit's owner. His son, Lindsay, stands to the far right. Fannie Mae Howard stands to the far left with Bing Crosby at her shoulder. The Seabiscuit/Ligaroti race was later immortalized in Laura Hillenbrand's award winning book *Seabiscuit, an American Legend*, which was made into a movie in 2003. (Courtesy Del Mar Thoroughbred Club.)

It was widely agreed among horse trainers that seawater had healing properties for the racehorses' legs; the cool, briny ocean at Del Mar became a regular venue for racetrack thoroughbreds. A photographer captured Charlie Whittingham, a trainer from Caliente who attended the first race meet and stayed until his death in 1999, engaging in this practice. (Courtesy Bill Arballo.)

The management team of the Del Mar Turf Club in the 1940s included Bing Crosby (center), Pat O'Brien as vice president (left), and Willard F. Tunney as general manager (right). Together they served to revitalize Thoroughbred racing in Del Mar as the World War II conflict came to its conclusion. The track reopened on July 11, 1945, for a 40-day meeting. (Courtesy Del Mar Thoroughbred Club.)

Bing Crosby poses with actress Betty Grable, who presented roses to the winning jockey during Del Mar's third racing season. Grable became a regular at the racetrack and at the Hotel Del Mar, where she lived during the summer. She was Hollywood's top draw in 1943 and was reported to be the highest-paid woman in the United States. A famed pin-up girl during World War II, her legs was insured for a quarter of a million dollars. (Courtesy Del Mar Thoroughbred Club.)

Jockeys who experienced both the physical and emotional stress of competing professionally in Thoroughbred racing were able to partially stave off some of the stress by playing softball. These games were less about skill and more for fun. Bing Crosby regularly participated in these games, helping to set a relaxed mood, one in which everyone had a tremendous time. (Courtesy Del Mar Thoroughbred Club.)

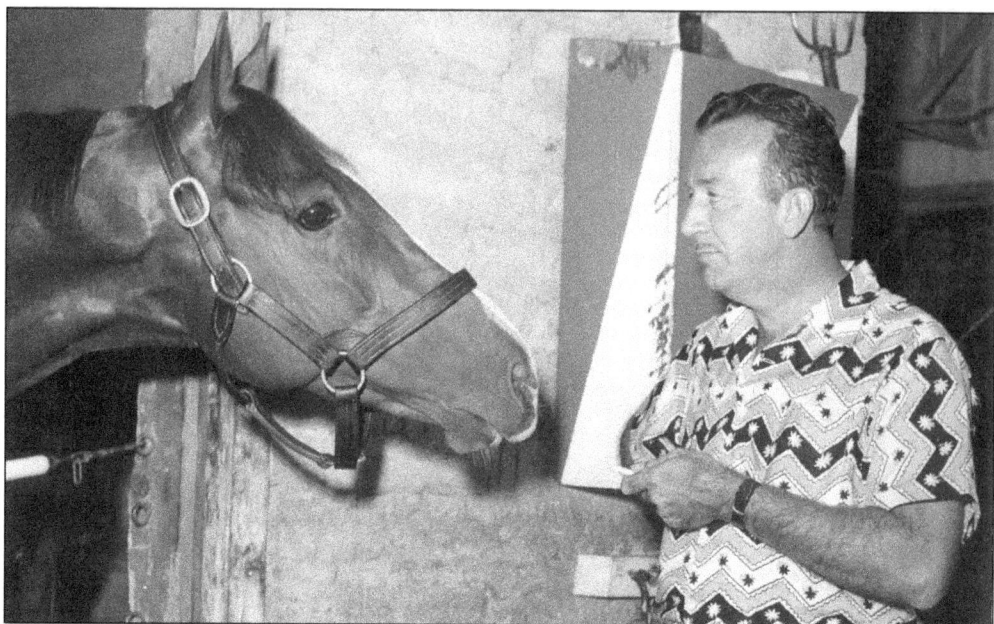

Racetrack celebrity sightings became a constant, especially as the new, hopeful decade of the 1950s emerged from the earlier, war-stricken 1940s and as stars began investing in Thoroughbreds. Famed bandleader Harry James is photographed giving a snack to his winning horse, Big Noise. The colt was the first stakes winner sired in America by Khaled and the winner of the 1951 Del Mar Futurity. (Courtesy Del Mar Thoroughbred Club.)

In the 1950s, FBI director J. Edgar Hoover used time after his annual physical at La Jolla's Scripps Hospital to visit the track. Here he poses (center) with jockeys (from left to right) Donal Bowcutt, Billy Pearson, Henry Moreno, Pete Moreno, John Longdon, Bill Shoemaker, R. L. Biard, Euclid Le Blanc, and Gordon Glissen. In the second row behind and to the left of Hoover stands Bert Thompson of the Jockey's Guild, and to the right is O. L. McKinney, general manager. (Courtesy Del Mar Thoroughbred Club.)

In 1948, Herbert Trent won the inaugural Del Mar Futurity with Star Fiddle, a gelding who 10 years later and until 1972 would lead the Futurity post parade with Trent in the saddle. Star Fiddle became a much-loved racetrack fixture, and when he died in 1978, the 32-year-old was buried in the Del Mar infield. (Courtesy Del Mar Thoroughbred Club.)

A historic milestone for Del Mar occurred on Labor Day in 1956, when jockey John Longden became the world's winningest rider, overtaking the record previously held by English champion Sir Gordon Richards. Longden, photographed here a nose ahead of his rival, celebrated his 4,871st win that day. (Courtesy Del Mar Thoroughbred Club.)

Fourteen years later to the day, celebrated jockey Bill Shoemaker would inherit the title of winningest jockey, shown riding a filly named Dares J. to his 6,033rd victory. Shoemaker would record another 2,800 wins before his retirement 20 years later in 1990. (Courtesy Del Mar Thoroughbred Club.)

The annual Del Mar Horsepark's yearling sale produced such remarkable winners as Officer, captured here taking home first place in the 2001 Del Mar Futurity. Called "one of the most brilliant two-year-olds of recent years" in *Del Mar, Its Life and Good Times*, Officer was the son of Bertrando, who had won the same title 10 years earlier. (Courtesy Del Mar Thoroughbred Club.)

Another recent Del Mar legend is Cigar, third from left, who was set to break Citation's mark of 16 consecutive wins. In 1996, Dare and Go, second from left, blazed past Cigar on the final stretch of the Pacific Classic, upsetting Cigar's anticipated victory. (Courtesy Del Mar Thoroughbred Club.)

Smiles were plentiful for the winner circle trophy holders who celebrated Juddmonte Farm's victorious Skimming, who had just won the 2000 Pacific Classic. Joe Harper, president of the Del Mar Thoroughbred Club (left), celebrates with jockey Garrett Gomez (center) and trainer Bobby Frankel. (Courtesy Del Mar Thoroughbred Club.)

Two star Del Mar jockeys exchange congratulations. A young Bill Shoemaker (left) with an astounding 52 victories following his maiden race a year earlier, and veteran champion John Longden (right) had deadlocked, each recording victories in 60 Del Mar races by the 1950 racing season. (Courtesy Del Mar Thoroughbred Club.)

Four

Exhibits,
Entertainment, and
Events

With the arrival of the 1950s, designers introduced more playful fashion, represented here by the polka dot. The polka dot had debuted as a design in the 1850s, coinciding with the popularity of the polka dance. Ivory/black was a favored combination. It reappeared in the 1950s as white/black polka dots on full skirts and soft blouses and in this example as a dance troupe's uniform of dinner-plate hats and tight-bodiced dresses. (Courtesy 22nd DAA.)

The 1946 fair was a cause for countywide jubilations, being the ideal catalyst to stage the celebrations of peacetime. "The Ferris wheel turns again," touted news reports. Paul T. Mannen was elected president of the 22nd DAA, and part of his team were manager Ernest O. Hulick and Fred Heitfield, appointed director of special events. The fair set an all-time attendance record that year of 170,000. (Courtesy 22nd DAA.)

In the 1950s, a visit to the Gay Way also meant a ride on the Ferris wheel. The first Ferris wheel was designed by George W. Ferris in 1893 for the Chicago World's fair. His original wheel was considered an engineering wonder, with two 140-foot steel towers supporting the wheel connected by a 45-foot axle and a wheel section with a diameter of 250 feet and a circumference of 825 feet—these twin wheels were considerably smaller. (Courtesy 22nd DAA.)

Fairest of the Fair Queen Gloria O'Rear, left, and her attendant Pat King, runner up of the beauty pageant, ceremoniously rang the old mission bell in the 1946 fair season calling thousands of fair visitors to Fiesta Time. Part of this daily event was the popular variety show held on the grandstand stage, known then as the Fiestacade. The show attracted big acts and famed celebrities. The Spanish theme dominated this year's fair, with the renowned Tipica Band of Mexico City headlining the show. (Courtesy 22nd DAA.)

In later decades, the roller coaster surpassed the thrill of the Ferris wheel. La Marcus Adna Thompson is considered the father of the American roller coaster. Thompson's creation, named the Switchback Railway, opened in the spring of 1884 at Coney Island in New York and made the inventor hundreds of dollars per day; this was remarkable considering only a nickel per ride was charged. History has dubbed him the "Father of the Gravity Ride." (Courtesy 22nd DAA.)

Agriculture was the main attraction at the county fair during its early decades. This 1948 exhibit showcases a collection of state-of-the-art, heavy-duty agricultural machinery, part of the Cavalcade of Transportation staged that year. It was quite a contrast to the horse-drawn machinery used by farmers in the area just 30 years earlier. (Courtesy 22nd DAA.)

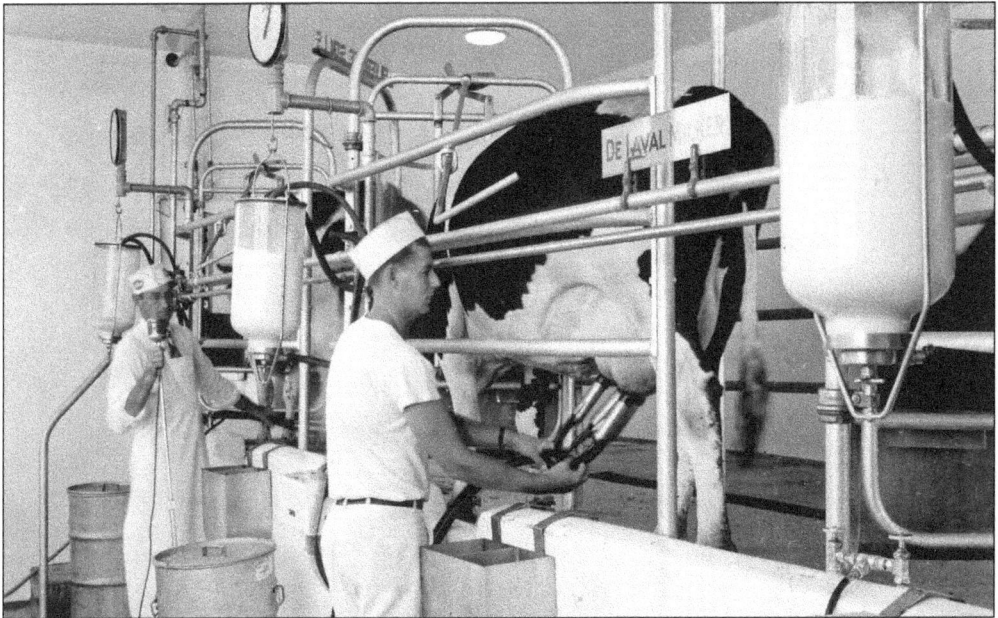

It was 1878 when Gustaf de Laval (1845–1913) got the patent on his invention, the cream separator, revolutionizing milk production. In 1917, the first De Laval vacuum-operated milking machine was introduced, and by 1951, the first herringbone parallel parlor was introduced with its automatic cleaning of pipelines. This invention drew many spectators during the 1951 fair. (Courtesy 22nd DAA.)

The exhibits and livestock contests entered by the region's 4-H Clubs were a huge contribution to the Junior Fair's events. Members of 4-H were eligible to participate from age 9 to 18. One of the rules was, and still is, that the entered animal must be clean, inoculated, and free of disease. Many slept with their livestock and, like this girl in 1950, read to her steers to pass the time. (Courtesy 22nd DAA.)

The 4-H Club, which stands for head, heart, hands, and health, required members to participate in group projects creating educational displays that were then entered at the county fair. Shown here are the 1951 winning entrants, the La Mesa Farmerteers, with their display that encouraged conservation. (Courtesy 22nd DAA.)

These penned hogs represent the importance of the farming community's contributions, and livestock judging was a major part of the fair's daily events, kicking off at 10:00 a.m. Categories included sows and boars in each of the following: two years and older, senior yearling, junior yearling, senior pig, junior pig, and championships. Groups and herds and barrows were also competitive categories. (Courtesy 22nd DAA.)

The classifications for judging rabbits was vast. Ranging from American blues and whites, checkered giants, Angoras, Belgians, beverens, chinchillas, Dutch, English, Flemish, Havanas, Himalayans, Japanese, lops, lilacs, Polish, New Zealands, Rhinelanders, Sitkas, and silvers, the judges must have had a tough task choosing their winning bunnies. (Courtesy 22nd DAA.)

Keri Shelstead of El Camino 4-H Club is proud to showcase her 1978 winning grand champion—a 226-pound market swine decorated with dozens of daisies and evidently named Paul for local famed floriculture king Paul Ecke Jr., who good-naturedly poses with her for this photograph. In memory of Ecke's passing in 2003, the fair's annual flower show is now named in his honor. (Courtesy 22nd DAA.)

"A comely 4-H member proudly displays some of the work done by members of the Future Farmers of America;" so noted the caption on this 1946 promotional photograph that appeared in area newspapers. "Colorful and smart summer frocks were made using fabric from sugar sacks," it went on. Sugar sacks were a necessity during wartime, when supplies were hard to come by. (Courtesy 22nd DAA.)

"Fifty pounds of good live pork," noted the press, "free to Paul de Lucia of Granger St. in National City." He caught his winning prize, a greased pig, at the 1946 fair. There were about 200 who participated in this dash around the racetrack enclosure. Part of the fair's attraction for kids was the many contests organized for them. These special events were under the direction of Larry King. (Courtesy 22nd DAA.)

These ladies are modeling their trophy-winning hat designs in the crazy hat contest, an event organized for the 1952 fair as part of the home arts exhibits. Creative use of breads and fresh fruits; cheese wheels and fake mice; and berry baskets with a toy monkey evidently caught the judges' eyes. (Courtesy 22nd DAA.)

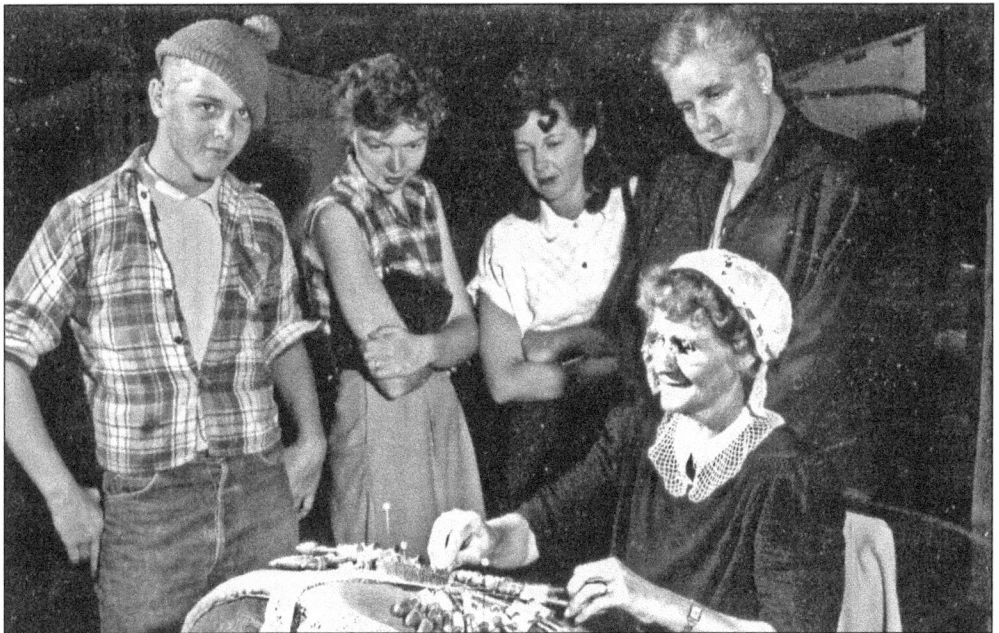

The delicate yet complex art of bobbin tatting is being demonstrated in this 1951 photograph, which captured a skilled demonstrator who was part of the home arts exhibits. She fashions dozens of small, wooden bobbins, wound with thin silken yarn, which are interwoven to produce intricate lace designs often used for collars, doilies, caps, and embellishments for clothing. (Courtesy 22nd DAA.)

A "hot dog for a hot dog" noted local papers who published this 1951 photograph of a junior fairgoer and her dachshund, an entrant for the mutt contest. This was the third year of the contest, which encouraged children to enter. Categories included the dog with the shortest or longest ears, the shortest or longest tail, the funniest dressed, the most intelligent, the tallest or shortest, the most obedient, and best of show. (Courtesy 22nd DAA.)

One of the contributors of the fair was the Model Airplane Association, who demonstrated their radio-controlled model winged crafts in a 1951 exhibition. A family event, even young children were able to participate. This little girl, captured by the fairgrounds' photographer, appears very possessive of her model plane equipped with a small but powerful engine. (Courtesy 22nd DAA.)

The iconic clown has been a perpetual attraction at the fair throughout its existence. While some very young children fear clowns, others welcome them, and clowns often appeared in Kiddie Land or as part of a circus event. These twins at the 1953 fair appear to be comfortable with their clown serenade. (Courtesy 22nd DAA.)

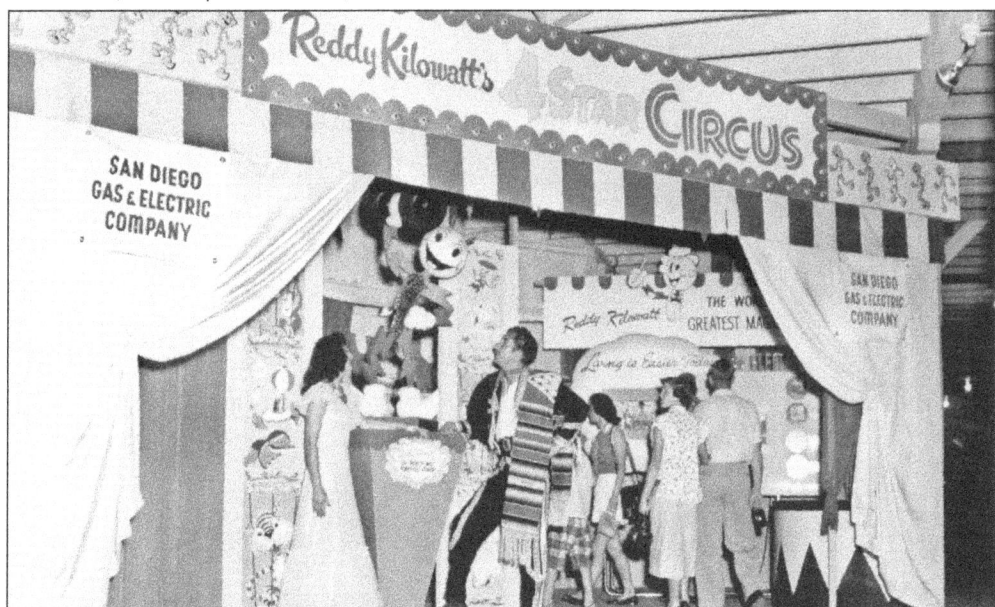

After World War I, Vorticism celebrated the sleek lines of modern technology. This movement later affected advertising, and a classic advertising icon created during this period was Ashton B. Collins's "Reddy Kilowatt." Inspired by a violent thunderstorm, he imagined a character that would be made of lightning. This character was used in the San Diego Gas and Electric Company's marketing, presenting the children's circus at the fair in 1951. (Courtesy 22nd DAA.)

As a family-oriented attraction, thousands of area children cut their teeth attending the summer fairs. To help if a child was separated from an adult, the Lost Child Car would provide a safe haven while searching for the missing parties. Sponsored by Dr. Ross's brand of pet food, the car was equipped with a huge bullhorn that could call out and help locate the separated adults. (Courtesy 22nd DAA.)

"Hold onto your three cornered pants kids!" wrote newspapers in 1950. Then called the "diaper derby," this entertaining event was strictly for babies 12 months of age and under. Comprising a 15-foot course, the race had one rule: the baby must crawl the entire distance. "Parents may stand at the finish line and call to their offspring," noted reporters, but there was to be no coaching from the sidelines. (Courtesy 22nd DAA.)

Del Mar's Ted Terwilliger was fairgrounds supervisor for a lengthy 30-year term. He was originally hired by manager Paul T. Mannen in 1947. Terwilliger's view of the fair was "from the inside, from the ground up and from beneath the surface," he said. When Ted retired in 1977, he said he was looking forward to visiting the fair with his son Don, "without worrying about lights going out and sewage overflowing." (Courtesy Don Terwilliger.)

Fair spectators thrilled at the roar of four vintage automobiles as they sped the 10-mile course around the racetrack in 1948, actually only reaching 60 miles an hour according to news reports but seeming to travel much faster. Famed racing driver Ralph DePalma flagged the winner, the No. 12 car, a Stutz. The No. 6 car, a Simplex, was close runner up. (Courtesy 22nd DAA.)

This airborne automobile illustrates how a trend at the fair—to entertain spectators with heart-stopping stunts—continued from year to year, a thrill that remains today. The souped-up 1951 model Ford must have been pretty robust to accelerate to such a velocity, become airborne, and clear the gap between the ramps. (Courtesy 22nd DAA.)

This 1951 automobile exhibition, staged in a big-top tent, spotlights the Studebaker Commander (far right). A bullet nose, single headlight, V8-configured engine dubbed the "Miracle Ride," it was then considered a new miracle of comfort and way ahead of its time. The series had a 12-year run from postwar 1947 through 1959. (Courtesy 22nd DAA.)

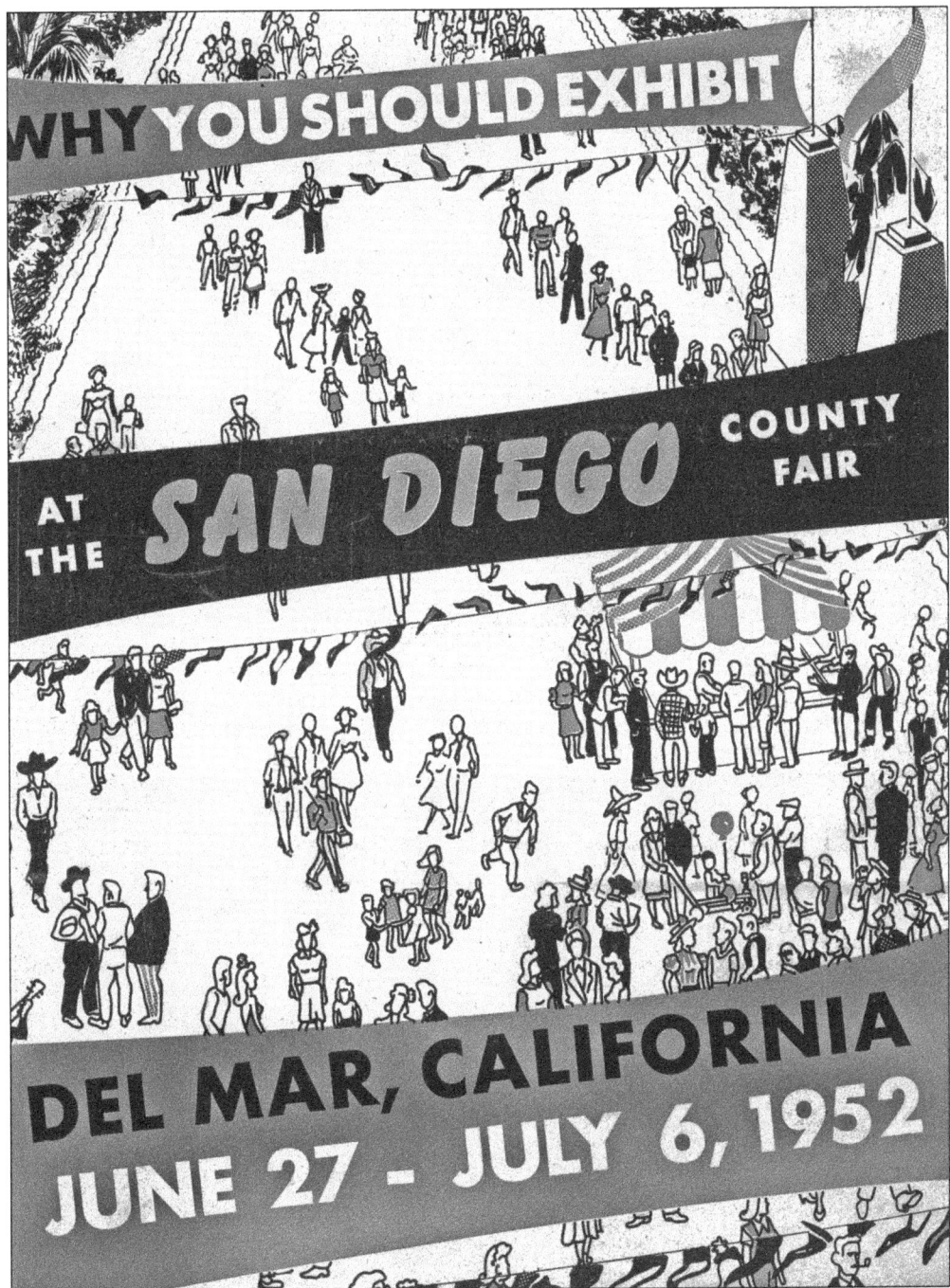

The fair's 1952 brochure contained important marketing statistics to attract future exhibitors and advertisers. For example, gross attendance for the prior year's fair was 202,000, and the average family income was noted as $3,573 per annum. "When you exhibit at this Fair you reach the prosperous middle brackets who buy the majority of the products for sale!" boasted the brochure. Exhibits included the Home Show, Auto Show, Miners and Gems Show, Farm Machinery Show, Flower Show, Hobby Show, Electrical Appliance Show, and Bazaar. (Courtesy 22nd DAA.)

Famed award-winning entertainer Fred Astaire is generally acknowledged to have been the most influential dancer in the history of film and television musicals and was rated the greatest dancer of the 20th century. The American Film Institute named him the fifth Greatest Male Star of All Time. He's shown here in 1946 with his son, Fred Jr., taking aim at the fair's shooting booth in the Midway. (Courtesy 22nd DAA.)

Another major attraction was the Gem and Minerals Show, which remains a sparkling addition to the fair exhibition events today. In this 1951 shot, an expert cutter, looked on by fascinated observers, is demonstrating the first step in how to turn a rough, uncut gemstone into a shining jewel. (Courtesy 22nd DAA.)

"What an eyeful! The Eiffel Tower out of toothpicks!" noted news writers who reported on the winning entrants in this special category of hobbyists. Several entrants competed for ribbons in this 1951 contest, which included some remarkable model ships, bridges, and buildings painstakingly constructed from thousands of wooden toothpicks. (Courtesy 22nd DAA.)

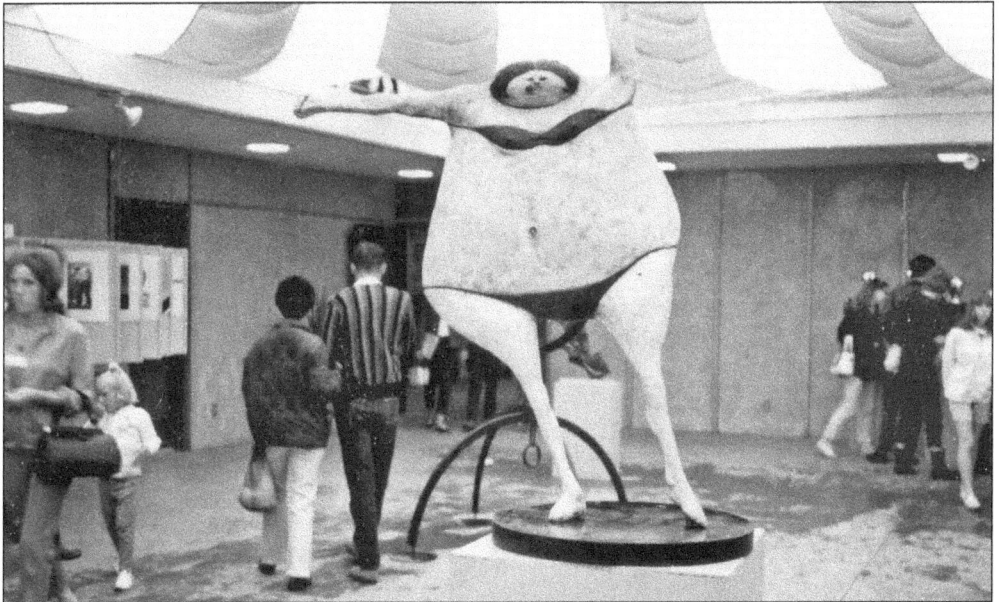

A much-anticipated fair exhibition was the Art Show, which over the decades expanded to include many categories. As new technology was developed, new art forms emerged, like photography, or more recently the digital arts. This whimsical bikini-clad sculpture welcomed spectators to the 1951 art exhibition. By the 1970s, the show was renamed the All Media Art Show and averaged 800 entries in all. (Courtesy 22nd DAA.)

Many consider the Flower Show to be the most popular of all the fair exhibits. Premiums for this department were at $3,000 in 1946. The *San Dieguito Citizen* wrote that year, "highlighting the flower show will be begonias, glads, carnations, dahlias, landscapes, rock gardens, lath houses, and nursery." To accommodate increased categories, the exhibition space was expanded by a third that season. Years later, the O'Brien Hall would house the Flower Show. (Courtesy 22nd DAA.)

Milton Sessions, nephew of Kate Sessions, hosted the Flower Show in 1993. In the 1920s and 1930s, Sessions landscaped many of the area homes, working closely with architect Richard Requa. In 1993, a palm tree was named for Sessions at the San Diego Zoo; he was also named horticulturist of the year by Cuyamaca College Botanical Society and was guest of honor at the Don Diego Banquet that year. (Courtesy 22nd DAA.)

The skilled art of floral arranging has long been a highlight of the Flower Show exhibition, part of the Floricultural Department's main events. Arrangements were prepared in a variety of categories—rose, dry arrangements, corsages, and arrangements by growers—as well as by type of flower—begonias, cut orchids, succulents, miniatures, cactus, rose, and others. (Courtesy 22nd DAA.)

In 1910, fifteen American florists began exchanging orders for out-of-town deliveries. Originally called Florists' Telegraph Delivery, FTD was the world's first flowers-by-wire service. FTD expanded to include international transactions 55 years later. Renamed Florists Transworld Delivery, it reflected its global presence. In 1914, FTD adopted the classic figure of Mercury as its logo. Today the "Mercury Man" remains one of the most recognized logos in the world. (Courtesy 22nd DAA.)

Kids wearing straight-legged blue jeans, plaid shirts, and canvas sneakers and chewing bubble gum was typical in the 1950s. But here it was for a good cause—the chance to win a much-coveted ribbon by competing in the Bubble Gum–Blowing Contest, one of the many light-hearted, fun events organized for junior fair contestants. (Courtesy 22nd DAA.)

Part of the attraction for many junior fair visitors was the opportunity to enter zany contests dreamed up by the special events department. In the 1950s, that meant competing in the Watermelon-Eating Contest. The winner was the fastest consumer of a large wedge of watermelon. There were no seedless varieties back then. (Courtesy 22nd DAA.)

Hot air balloons were a main attraction at the fair in 1970 and continue to be synonymous with the coastal area of Del Mar. A much-loved tourist attraction, the brilliantly colored balloons can be seen year round floating silently as the sun sets over the coastal cities and inland Rancho Santa Fe. (Courtesy 22nd DAA.)

In the 1960s, the space race was in many people's consciousness and was brought to its climax in 1969 when Neil Armstrong was the first man to walk on the moon. Six years earlier, San Marcos–based General Precision, Inc., founded in 1937 as a high-tech manufacturer of aircraft, missiles, computer components and space vehicles, had debuted its Moon Landing exhibit at the fair. (Courtesy 22nd DAA.)

It has long been a tradition for the summer-staged fairs to organize a spectacular fireworks display in front of the grandstand on the evening of July Fourth. In later years, as technology advanced, the displays became more elaborate, were synchronized to music, and thrilled crowds of thousands, which included area residents who would pack nearby streets to view the show from afar. (Courtesy 22nd DAA.)

Five

ARMED FORCES

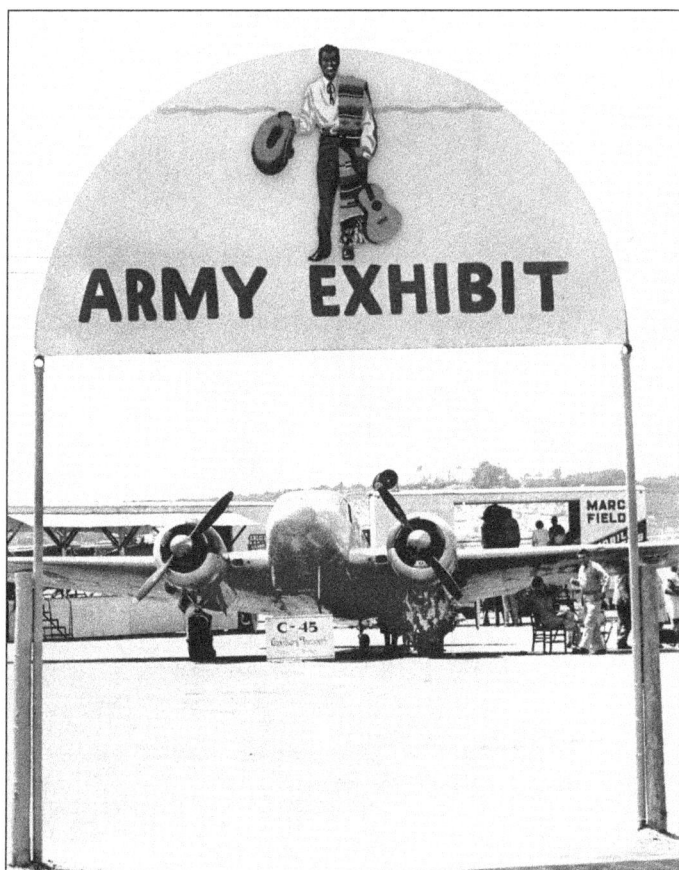

Don Diego stands ready in 1946 to welcome the very first military aircraft exhibition to the fairgrounds. Through an open arch, visitors view this vintage World War II Beech C-45 aircraft from March Field. In total, 5,204 C-45s, including other variations of the military Beech D-18 series, were delivered during the war. This durable and reliable design was produced from 1937 to 1969. (Courtesy 22nd DAA.)

On July 1, 1943, eighty acres just east of the Del Mar fairgrounds served as an airship landing area, which grew into this controlled airport. The U.S. Navy operated the facility along with its Santa Ana Air Station. Around 1946, the barracks, control tower, and active runway 27 are clearly visible, along with an array of aircraft of the period with a panoramic view of the surrounding area. (Courtesy 22nd DAA.)

Actor Wallace Beery (left) played Ned Trumpet, the pilot of a World War II blimp, in the 1944 movie *This Man's Navy*, which was shot at the U.S. Navy blimp base then located just east of the fairgrounds. Tony's Jacal restaurant and a worm castings business were housed in the former barracks there. (Courtesy Del Mar Historic Society [DMHS].)

In 1941, U.S. Marines from Camp Elliot trained at the Del Mar Racetrack, using the stables as facilities and the track for endurance exercises. Robert E. Hofferd (top center) would later perish at the Battle of Iwo Jima. Daniel P. Blankowski (left) died as a civilian in 1985. There are no later records for the third marine, William G. Donnelly. By September 1942, marines would be stationed at Camp Pendleton. (Courtesy 22nd DAA.)

In 1943, the fairgrounds were used to aid the war effort. Women like Del Mar's Marge Dunham (left) and Mary Marquez filled in the labor gap. Known in the local papers as "Bing's Bomber Builders," these real-life Rosie the Riveters manufactured subassembly ribs for the B-17 Flying Fortress bombers. Production was temporarily set up in the fair's grandstand building and managed by the formerly retired Fred Poggi. (Courtesy 22nd DAA.)

Army pilot Lt. Col. John Coleman Herbst, a World War II ace with 18 confirmed kills in China/Burma air operations, was killed in an air show maneuver at the fairgrounds on July 4, 1946. His diving P-80 jet fighter crashed, failing to pull up in time. Herbst's second wife of less than 24 hours and his 13-year-old son looked on. Few photographs of him remain extant. (Courtesy 22nd DAA.)

March Army Air Base flying team commander Lt. Col. John Coleman Herbst was 36 when he fatally crashed into a ditch executing final capability displays of the P-80 Shooting Star jet at the fairgrounds' air show in 1946. This news photograph captures wreckage strewn in mud just west of the fairgrounds. Herbst's wife and teenage son helped with the futile emergency rescue effort. (Courtesy 22nd DAA.)

Maj. Robin Olds, Army Airforce pilot (left), and Don Diego (right) escort Queen for a Day winner Bobbie Moore to the grandstand of the 1946 fair. Saluted by a detachment of U.S. Army soldiers, they march forth during the opening day ceremonies. Sadly that week, on July 4, Major Olds flew in an aerobatic exhibition resulting in the death of World War II flying ace John Coleman Herbst. (Courtesy 22nd DAA.)

Visitors view the cockpit of a Brewster F2A-3 Buffalo fighter, the first monoplane to serve on U.S. aircraft carriers. Carrying four 0.5-inch machine guns, it was first produced in 1939. Most were sent to Britain, as these markings show. The F-4U Corsair (center) is displayed with all 2,300 horsepower and a four-blade propeller towering above onlookers. It could fly at 470 miles per hour and sported four 20-mm cannons. (Courtesy 22nd DAA.)

The bugler tattoos the military colors as Vice Adm. J. B. Oldendorf (left), commandant of the 11th Naval District, and Maj. Gen. W. H. Hale, commander of the 4th Air Force, salute sharply in traditional display. Leo Carrillo (right), period film star and local celebrity, gives a stately civilian salute, hat over heart, and stands proudly next to the military VIPs in the opening ceremonies of the 1946 fair season. (Courtesy 22nd DAA.)

Underwater welding attracts a fair crowd in 1946 as divers from the U.S. Navy demonstrate their skill at this highly dangerous task. The tank at center allows viewers to observe a diver welding a sample project under the waves as two sailors above assist and maintain safety. A diving suit can be seen hanging from the railing. Boom lighting is provided for night exhibitions. (Courtesy 22nd DAA.)

The B-29 Superfortress fuselage becomes a tourist attraction at the 1951 fairgrounds. Lines formed to see the interior of the aircraft that "won the war" by dropping two atomic bombs over Japan in August 1945. An arms race between the powers of the world led to development of progressively larger strategic bombers. The B-29 was the largest. It subsequently led to the development of the B-52 bomber. (Courtesy 22nd DAA.)

Aerial views of the army/air force's 1951 exhibit also captured the popular snack concession (right center), which fed spectators waiting to view their weapons systems display. The B-29 Superfortress fuselage (bottom), the nose section of a B-17 Flying Fortress (center left), and the McDonnell XF-85 Goblin (upper right) were exhibited. The army also displayed its latest tank and truck, touted as the best in ground troop transport (upper center). (Courtesy 22nd DAA.)

The famous nose art (left) of World War II airmen is shown on the front of this B-29 Superfortress. Fair attendees in 1951 step up to peek into the cockpit of their aviation history. Lines of people of every age formed to view this fitting monument to aviation achievement provided by the U.S. Air Force. The majestic grandstand provides a proud backdrop to the exhibit. (Courtesy 22nd DAA.)

"Don't be a dummy—Join the Army!" Two uniformed mannequins stand guard over a display offering a prerecorded message about air force opportunities transmitted by telephone handset (center). This c. 1950 exhibit shows joint U.S. Army and U.S. Air Force recruiting posters lauding benefits of joining at age 18 and retiring at age 38 (far right). (Courtesy 22nd DAA.)

An army soldier encourages curious future recruits at sighting in the water-cooled .50-caliber heavy machine gun (center) while the .30-caliber light machine gun (right) goes unattended. The mortar (left) stand next to the tank without observer interest. A display of hand-held weapons (rear) gains a small crowd at the 1951 Del Mar Fair Army Weapons exhibit. (Courtesy 22nd DAA.)

The F7F night fighter is viewed by a couple as the pilot talks about it. The last propeller-driven World War II fighter built, all but a few orders for the F7F were cancelled by the military when Japan surrendered ending World War II. This rare, c. 1950 view of the F7F on exhibit at the Del Mar Fair by the U.S. Navy is unusual since only the U.S. Marine Corps ever flew these craft. (Courtesy 22nd DAA.)

With a model of "Old Ironsides" (center) displayed as a serious symbol of past U.S. Naval tradition, a first-class petty officer partakes of contemporary navy tradition: a bit of hula fun. These two unnamed sailors enjoy the 1953 Del Mar Fair's naval exhibit, showing models and artifacts of naval life to visitors by photographing the lighter side of fighting men. (Courtesy 22nd DAA.)

The 1963 fair displayed this Martin Mace (designated as TM-76 tactical missile until 1963, then as MGM-13 for mobile-launched versions) tactical surface-to-surface missile. It was first deployed in 1956. Mace was launched using a solid-fuel booster rocket for initial acceleration and an Allison J33-A-41 turbojet for flight. The MGM-13 remained in service until the early 1970s. (Courtesy 22nd DAA.)

As the airman shows the marvels of jet air power to Fairest of the Fair Marla English, he points out the technical highlights within the cut-away sections of the Allison J33-A-9 engine. This engine was used in the earliest operational jet fighters, like the P-80 Shooting Star, which performed aerobatic demonstrations at the 1951 fair. (Courtesy 22nd DAA.)

The 1953 fair continued the tradition of military exhibitions, which drew larger and larger crowds. The Navy Diving Team shows their stuff by performing underwater demonstrations within this windowed tank to an engaged crowd of enthusiastic onlookers. Typical tasks performed might include underwater welding and other technical skills. Safety is stressed, and the sailor attending (left) assures all systems are go. (Courtesy 22nd DAA.)

Six

THE NATIONAL
HORSE SHOW

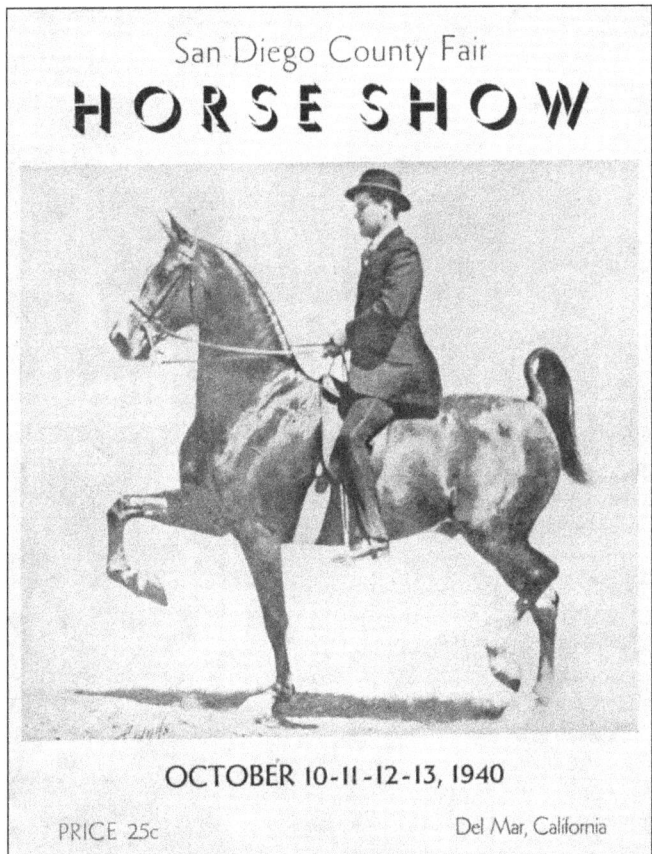

San Diego County Fair

HORSE SHOW

OCTOBER 10-11-12-13, 1940

PRICE 25c

Del Mar, California

In 1940, what was then known as the San Diego County Horse Show (not then having reached national status) had a four-day run in the fall. Directed by the 22nd DAA, with a horse show committee comprising six locals, first-prize cash awards were as high as $250 for the Hunters and Jumpers Championships and as low as $15 for Pinto Horses Under Saddle. The event was held a month and a year after the start of World War II in Europe. (Courtesy DMHS.)

CTION
National
MAY 2, 1946

The San

'Railbirds' Prejudge County's National Horse Show

With World War II over, the celebrated Betty McClanahan, president of the San Diego National Horse Show Association, led the movement to reestablish horse show enthusiasm by offering prizes adding up to $20,000 in 1946—nearly a quarter-million dollars in present values. It got headlines. A *Union Tribune* clipping shows committee members (from left to right) Justin Evenson, Parker Sietz, Fred Simpson, and Vroman Dorman reviewing the categories for prizes. (Courtesy DMHS.)

Fifty-eight pages covering 50 years of National Horse Show history lie beneath the cover of this souvenir program of the 1995 golden anniversary celebration. With dedications by then–California governor Pete Wilson and Del Mar Fairgrounds general manager Timothy J. Fennell, a comprehensive chronology of the show's history, including selected photographs of event action, are complete and memorable. Feature articles for 1995 covering event rules and conduct are included. (Courtesy DMHS.)

Progressing from the successful 1946 season, 1947 found burgeoning crowds of visitors. The National Horse Show attendance at the fairgrounds showed increased public interest in new entries, which spurred officials to add events like this bucking bronco competition. Numbers of total competitors increased from 350 in 1946 to over 800 in 1947, and prize money skyrocketed from $20,000 to $30,000 in that single year's time ($335,000 in present values). (Courtesy 22nd DAA.)

The year 1946 marked the first year that a cutting horse competition was created in Texas. The next year brought cutting horses and steer roping as well as other variations of contests from the American West to Del Mar. These exciting new contests were exhibited at the fairground's National Horse Show for the first time in 1947. This intrepid cowboy tackles a steer as he competes for the trophy in the inaugural competition in steer roping. (Courtesy 22nd DAA.)

San Diego

National Horse Show

Del Mar Fair Grounds

July 2 to 7, 1946

$20,000 in Prize Money

$1,000 Championship Stakes

**Full Classifications for Saddle Horses,
Roadsters, Hackneys, Harness Show Ponies,
Hunters, Jumpers, Stock Horses**

JUDGES: Douglas N. Davis, Lexington, Kentucky
Saddle Horses, Roadsters, Hackneys.

Tom Pilcher, Pacific Palisades
Hunters and Jumpers.

Gunner Thornberg, San Francisco
Stock Horses.

**Address All Communications to San Diego
National Horse Show Association, 301 Via Del Norte,
La Jolla, California**

This bulletin announces the establishment of the first official San Diego National Horse Show to be exhibited since World War II, which had disrupted event activity at the fairgrounds since 1941. Championships of $1,000 were large sums—equal to about $12,500 in present value. They attracted the best contestants and grew to be an internationally renowned equestrian event that seemed to grow ever more popular with visitors. (Courtesy 22nd DAA.)

STAR OF THE SHOW
DEL MAR NATIONAL HORSE SHOW

HAP HANSEN

The featured star of the Del Mar National Horse Show is shown here displayed on a full page in *Horses Magazine* in 1978. Hap Hansen was the most popular equestrian performing in the show. The magazine noted that the fairgrounds' 33rd annual event was "a big and successful show." The event has become ever more popular since then. (Courtesy DMHS.)

This 1960s shot provides a rare glimpse of the behind-the-scenes ring crew that enabled the National Horse Show's equestrian events to be staged. On the far right is the ring steward who bugled a fanfare signaling the start of each event, and to the far left is Robert "Chuckles" Hernandez. The announcer would often give public credit to the crew, declaring them "the fastest ring crew in the world!" (Courtesy Robert "Chuckles" Hernandez.)

Robert "Chuckles" Hernandez (right), a Solana Beach resident, was hired as seasonal staff by Manager Paul T. Mannen in 1953 to help maintain the grounds. Three years later, he became permanent, working for the Del Mar Turf Club at what would ultimately become a 40-year endeavor. In the summer season, he worked for the Turf Club during the day and for the National Horse Show at night, putting in 16-hour work days. (Courtesy Robert "Chuckles" Hernandez.)

Western technical show manager Larry Gimple rode a winner owned and trained by his parents, Pat and Glenn Gimple, at the Del Mar National Horse Show held at the fairgrounds in 1964. Having taken first in 10 and under, the win launched his venerable career in the horse show business, where he is held in the highest esteem by the industry. (Courtesy DMHS.)

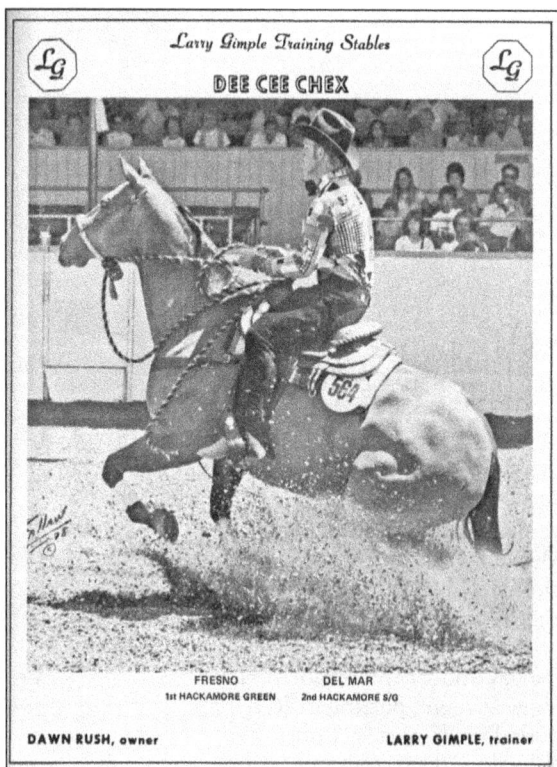

Larry Gimple Training Stables

DEE CEE CHEX

FRESNO
1st HACKAMORE GREEN

DEL MAR
2nd HACKAMORE S/G

DAWN RUSH, owner

LARRY GIMPLE, trainer

Horse number 564 is the champion Dee Cee Chex owned by Dawn Rush and trained by Larry Gimple. This 1965 action shot, depicting Glen Gimple's skills in reining and equitation, was displayed in a full-page lay-up in *Horses Magazine*. The Glen Gimple Memorial Bronze Trophy was established after this horse won the blue ribbon in Western Pleasure. (Courtesy DMHS.)

Di Ann Lundy of Los Angeles is pictured winning the Santa Anita Perpetual Trophy for leading rider awarded in 1975. She is depicted riding Danny in the Del Mar Fairgrounds National Horse Show to win as she displays the Santa Anita Sash and jumps a triple bar. She won here for the second year in a row. (Courtesy DMHS.)

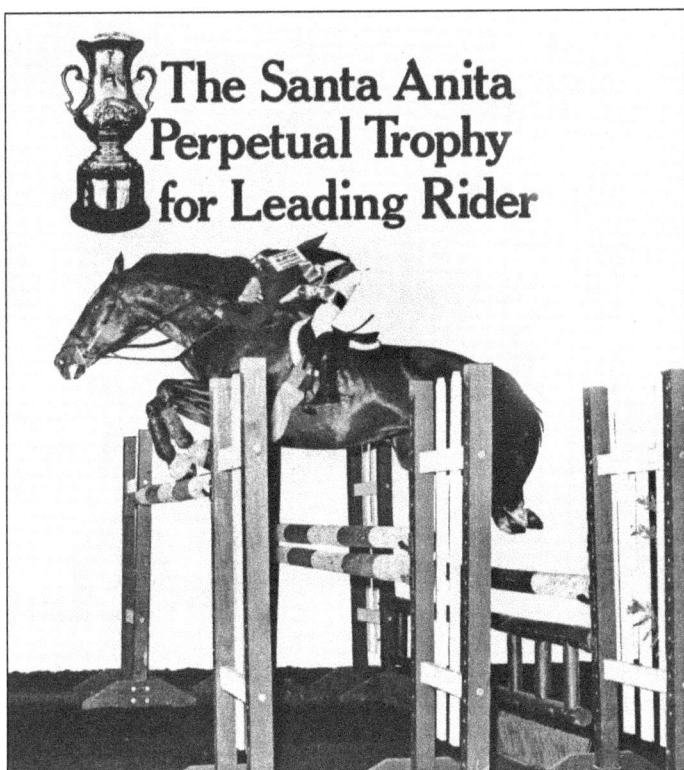

The Santa Anita Perpetual Trophy for Leading Rider

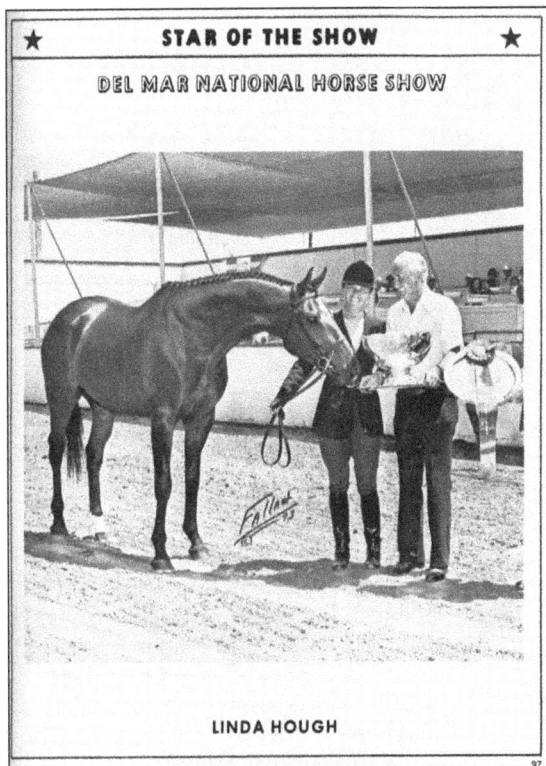

★ STAR OF THE SHOW ★

DEL MAR NATIONAL HORSE SHOW

LINDA HOUGH

Jimmy Williams presents Linda Hough with the Grand Prize Trophy for winning the Hunter competition in the Del Mar Fairgrounds National Horse Show in 1973. Manager Alan Balch produced the show that year in a flawless exhibition of 1,294 horses. Linda Hough was said to have had "countless excellent rounds" and was lauded as the "Star of Show," as noted in the National Horse Show program. (Courtesy DMHS.)

The finishing touches are completed in 1990 by masons working on the concession building to support the new, 4,000-seat, $5.4-million arena complex. The arena was needed to support the ever-larger crowds of fans to the fairgrounds' National Horse Show. In 1993, the visitor count exceeded 40,000 for the first time. (Courtesy DMHS.)

Jane Gardner smiles at Ace Command, the horse she owned and rode in the 1946 Del Mar Fairgrounds National Horse Show. Bought from Daneshall Stable in Louisville, Kentucky, this horse was the winner of the National Futurity as a three-year-old. The pride of their winning spirit beams from both rider and horse as they proudly saunter to the stable. (Courtesy DMHS.)

Barbara Worth Zimmerman, contestant No. 40 out of Barbara Worth Stable in Sacramento, California, clears the final hurdle to the trophy as she wins the Del Mar Fairgrounds National Horse Show high-jump competition in 1946. Prizes in this competition could be as high as $1,000, and the competition for this prize was fierce. (Courtesy DMHS.)

Proud owner Dale Harvey stands with Fast Lane as the 1984 winner of the 49th annual Hunter competition. This promotional photograph was taken at the completion of the competition. This was a year that found expanded competition from Japan, Mexico, Canada, and other countries, which raised a significant challenge to the trophy. Many contestants were training for the Olympic games at the Del Mar National Horse Show. (Courtesy DMHS.)

In 2005, event officials adopted the logo for the Del Mar National Horse Show. It now serves as a colorful and readily recognizable international symbol of this globally revered equestrian exhibition. People from the world over come to see the events at the fairgrounds every year. (Courtesy DMHS.)

Del Mar Horsepark spans a 65-acre, world-class equestrian facility located three miles east of the Del Mar Fairgrounds. It provides full-service accommodations for equestrian events of all levels. It has on-site catering, equipment rental, security, RV hook-ups, and plenty of parking. The facility has permanent box and pipe stalls for year-round boarding and offers professional training from beginning to Olympian levels. (Courtesy DMHS)

Seven

TODAY'S FAIRGROUNDS

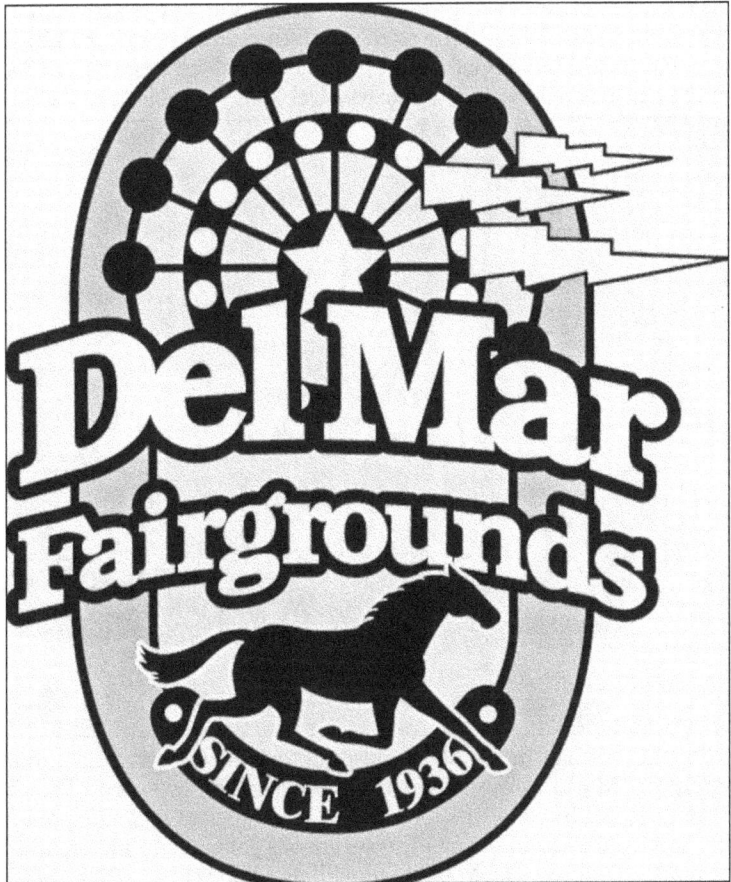

Today more than ever, the Del Mar Fairgrounds represents a valuable community resource as well as a first-class facility. Its mission statement, "To manage and promote a world class, multi-use, public assembly facility with an emphasis on agriculture, education, entertainment, and recreation in a fiscally sound and environmentally conscientious manner for the benefit of all," continues to be the benchmark by which it diligently conducts its business. (Courtesy 22nd DAA.)

SEPTEMBER 16, 1991

As the 1991 fair and racing season closed, wrecking crews arrived to facilitate the long overdue, $80-million remodeling of the grandstand structure. The racetrack's general manager, Joe Harper, left, and then-director Harry (Bud) Brubaker took ceremonial first swings symbolizing the start of the construction project. (Courtesy Del Mar Thoroughbred Club.)

With the west side of the old grandstand torn down, work began immediately on construction. Architect Morio Kow, whose firm had designed Hollywood Park and redone Keeneland in Kentucky, was hired for the job. He recommended a two-phase project. Award-winning Centex Golden Construction Company was hired as general contractor, and by the next season, racing resumed. (Courtesy Del Mar Thoroughbred Club.)

110

Following the 1992 season, construction work continued on the grandstand remodel, razing the remaining half of the structure. By the opening of the 1993 meet, the project was complete. The integrity of the Mission Revival style of architecture had been maintained, and the building was deemed a success. Gov. Pete Wilson was in attendance for opening day to help celebrate the occasion. (Courtesy Del Mar Thoroughbred Club.)

The Don Diego Clock Tower continues to be a central vantage point in the fairgrounds' midway. This recent photograph captures the 22nd DAA offices just behind the clock tower and the racetrack grandstand just beyond that. Since Sam Hamill's initial concept in 1936 of a Mission Revival–style complex, the fairgrounds have retained their regional architectural style. (Courtesy 22nd DAA.)

The 22nd DAA is governed by a nine-member board, all appointed by the governor of the state of California. Each serves a four-year term. The board meets on the second Tuesday of every month except July. As of January 24, 2008, the current board members are, as shown from bottom left to right, (first row) Vivian Hardage, Kim Fletcher, and Ruben Barrales; (second row) Michael Alpert, Pres. Russell Penniman, and Kelly Burt; (third row) Ann Davies, Barry Nussbaum, and Doug Barnhart. (Courtesy 22nd DAA.)

Tim Fennell has served as the chief executive officer and general manager for the 22nd District Agricultural Association since March 1, 1993. Since then, Fennell and his faithful team have increased annual fairground events to 300, improved its ranking to the fifth largest in the United States, and increased annual revenues to nearly $60 million. Fennell is currently chairman for the San Diego North Convention and Visitors Bureau, secretary of the Del Mar Racetrack Authority, and president of the Don Diego Fund. When asked about the county fair, he said, "I think of it as a 22 day long party for 1,250,000 of my closest friends with thrill rides, entertainment, exhibits, livestock and unique food delights." (Courtesy 22nd DAA.)

The Del Mar Thoroughbred Club's director, president, general manager, and chief executive officer Joe "The Cowboy" Harper can be observed riding horseback during his daily rounds of the Del Mar Racetrack's facilities. The grandson of famed director Cecil B. DeMille, Harper started as a photographer in 1966. By 1977, he had risen in the ranks and become manager, a 30-year commitment, leading the Del Mar track to unparalleled growth. (Courtesy Del Mar Thoroughbred Club.)

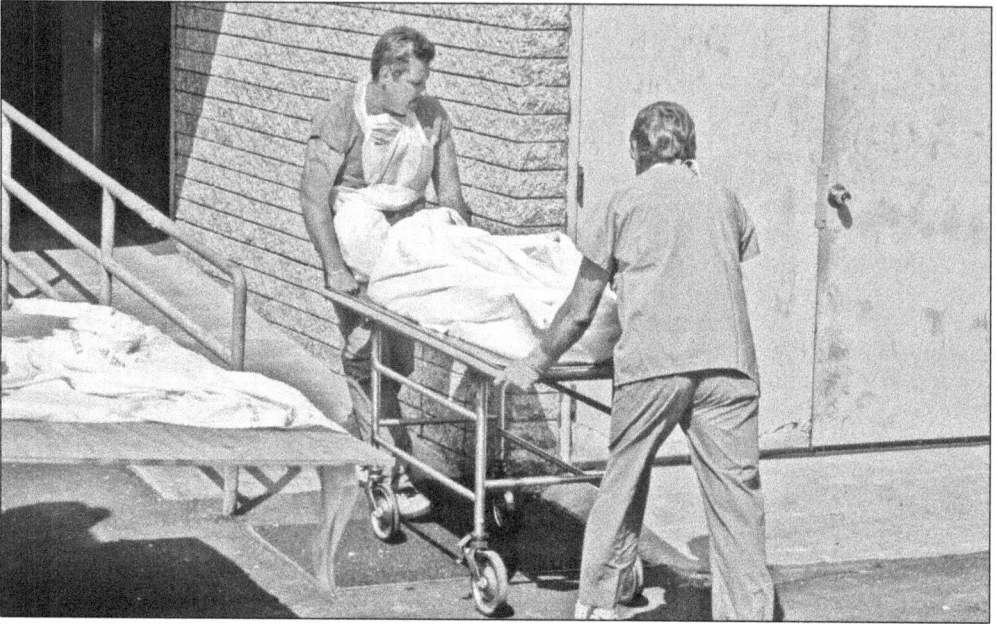

In 1997, the Del Mar Fairgrounds were used to accommodate reporters and photographers as the investigation into the aftermath of a local cult's bizarre philosophies got underway. Members of the Heaven's Gate Cult believed that once free of their bodies, they would be whisked to paradise by spaceship. On March 26, 1997, each member was found dead in the cult's Rancho Santa Fe mansion. (Photograph by Tom Keck; courtesy Tom Keck.)

As the body count in the Heaven's Gate investigation reached 39, authorities realized that this was a mass suicide of an unprecedented level. The media were caught up in the frenzy caused by the breaking news of the Rancho Santa Fe–based cult members who followed leader Marshall Applewhite to their deaths. The 22nd DAA played host to hundreds of reporters, many from overseas, who reported on and televised the proceedings. (Photograph by Tom Keck; courtesy Tom Keck.)

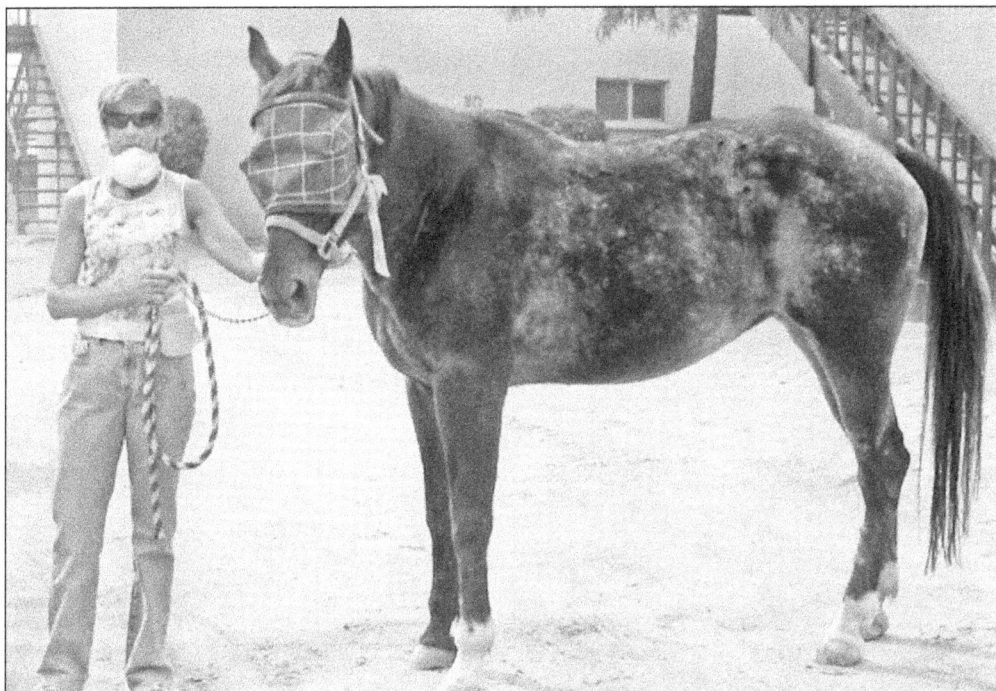

When the first wisps of wildfire smoke that would ultimately go down in history as "Firestorm 2007" were reported on Sunday, October 21, the 22nd DAA team reported to work knowing that the Del Mar Horsepark and the fairgrounds would repeat their role undertaken during the 2003 Cedar Fire. Once again these facilities served neighbors as an evacuation center. (Courtesy 22nd DAA.)

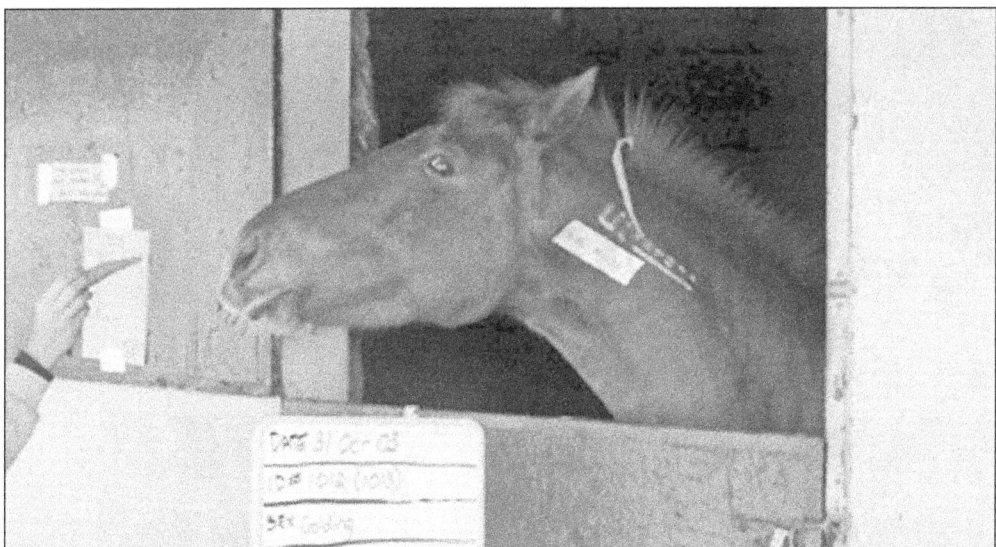

Tracking animals and matching them with owners was a major concern for the fairgrounds personnel. A checks-and-balance system was put in place, with each animal being given a unique number that was written directly on its neck or upper body. The number corresponded with that on a chart that also noted the name, sex, color, distinctive markings, ownership, and date of arrival. The firestorms of 2007 left thousands of animals homeless. (Courtesy 22nd DAA.)

The media were quick to respond to the emergency situation. Local television media were on-site to report the evacuation efforts and to provide up-to-the-minute updates for concerned citizens. Televisions strategically placed in the exhibition halls where evacuees were stationed helped inform the displaced families of the fire's destruction. (Courtesy 22nd DAA.)

Using the stables adjacent to the racetrack as refuge during the firestorms of 2007, these donkeys displaced from a local ranch look content in their companionship at the fairgrounds. Within a matter of hours of being declared an evacuation shelter, the Del Mar Fairgrounds team had collected 1,300 bales of hay, 5,000 bales of shavings for animals, and 2,000 cots and 2,500 pillows and blankets for people. (Courtesy 22nd DAA.)

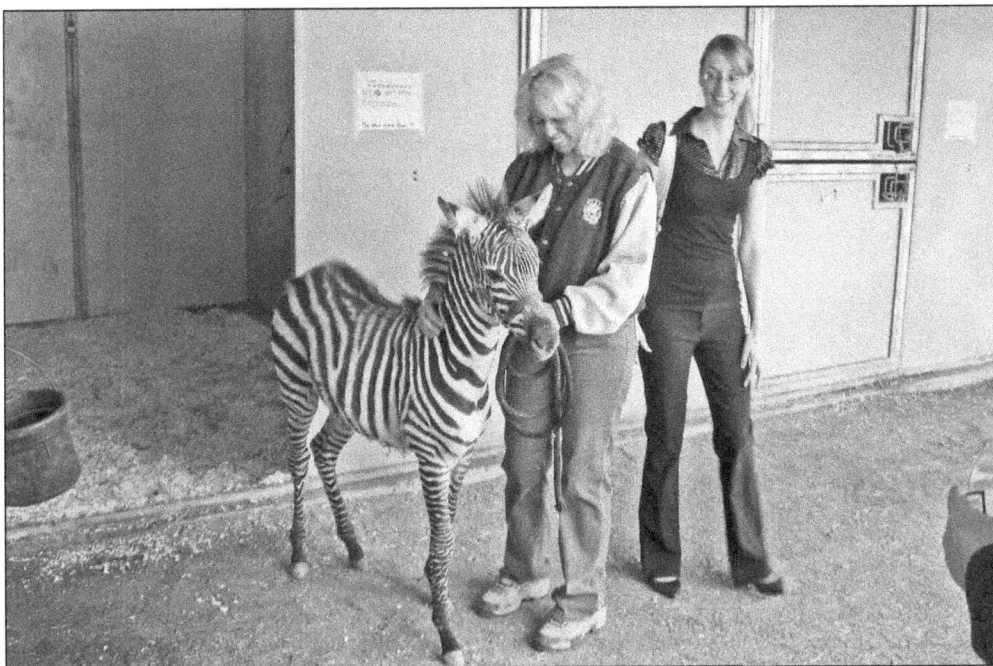

The disaster led to a mammoth collaboration of 22nd DAA personnel and willing volunteers. More than 1,000 individuals came forth to assist. There were so many volunteers that some were turned away. The Del Mar Fairgrounds and the nearby Del Mar Horsepark served as safe havens for almost 3,000 evacuated animals. Horses, donkeys, llamas, goats, cats, birds, six zebras, and a zorse (a hybrid horse and zebra) found shelter there. (Courtesy 22nd DAA.)

On Sunday, November 18, 2007, the Del Mar Fairgrounds and the Del Mar Horsepark ran a full-page advertisement in the *San Diego Union Tribune* thanking all those who had aided in their facilities becoming a safe refuge during the fire disaster. A photographic montage included several images of the National Guard, who had helped greatly during the six days of disaster relief, providing a sense of security during uncertain times. (Courtesy 22nd DAA.)

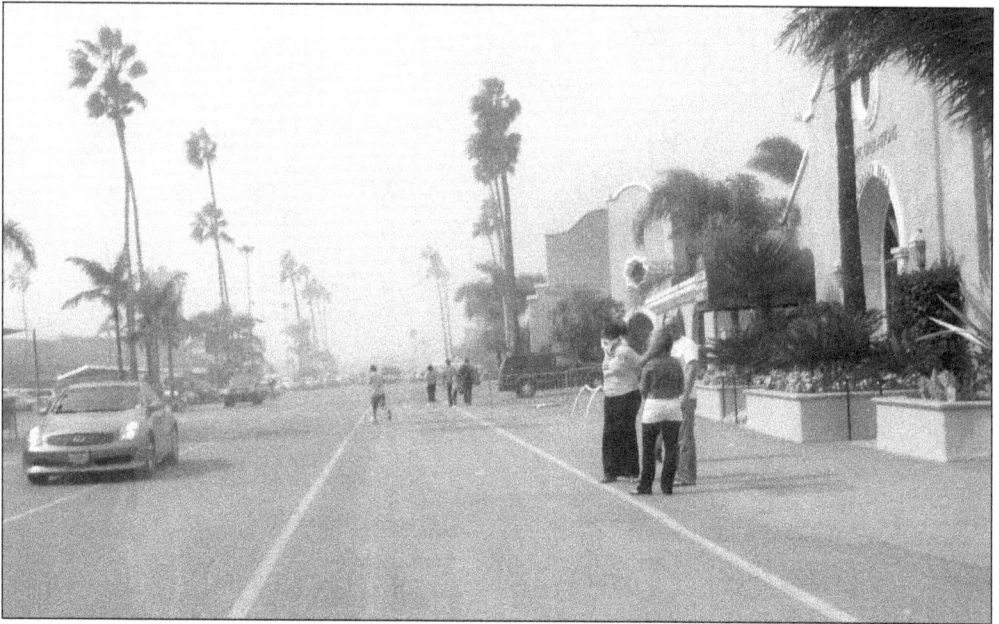

People and animals fell victim to the firestorm, which began as a brush fire fueled by seasonal Santa Ana conditions—extremely dry westerly winds. It ultimately scorched 300,000 acres, saw over half a million people evacuated, and became the largest ever evacuation effort in California state history. As people scrambled to protect pets and pack belongings, the fairgrounds provided calm amid the chaos. (Courtesy 22nd DAA.)

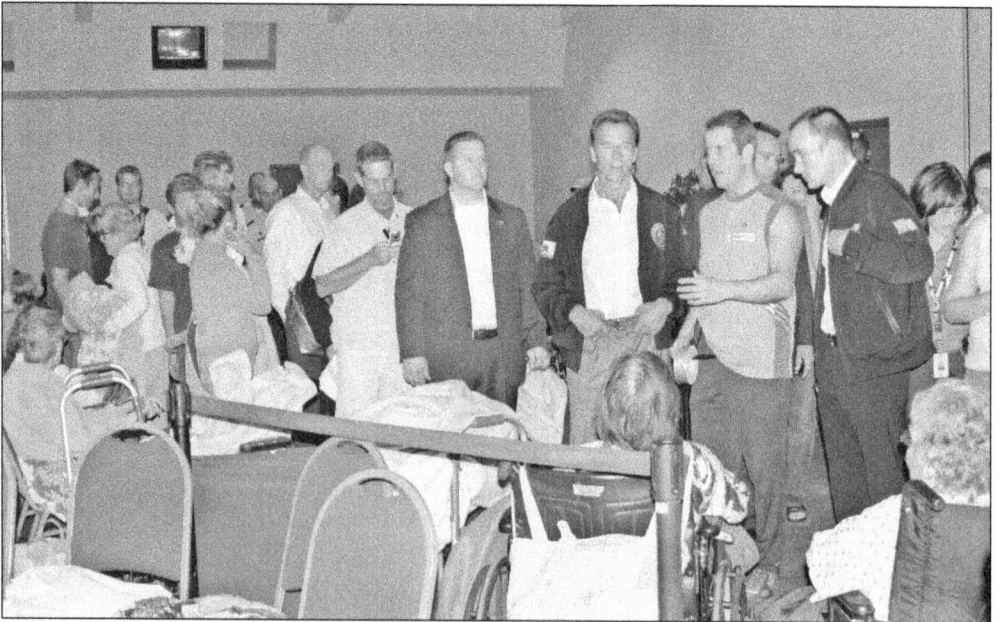

California governor Arnold Schwarzenegger was on the scene to survey first hand the incredible outpouring of relief efforts. With such short notice, it was nothing short of miraculous how the 22nd DAA, its food and beverage partner Premier Food Services, and numerous local companies donated items to aid in the comfort and nourishment of 2,200 evacuees and 3,000 animals. (Courtesy 22nd DAA.)

When the fledgling years of the fall fairs, originally organized as autumnal harvest produce showcases, gave way to summer festivals, Independence Day celebrations were incorporated into the county fair line-up of events. For many locals and tourists, attending the fair on July Fourth has become an annual tradition, with record attendances reflecting this trend. (Courtesy 22nd DAA.)

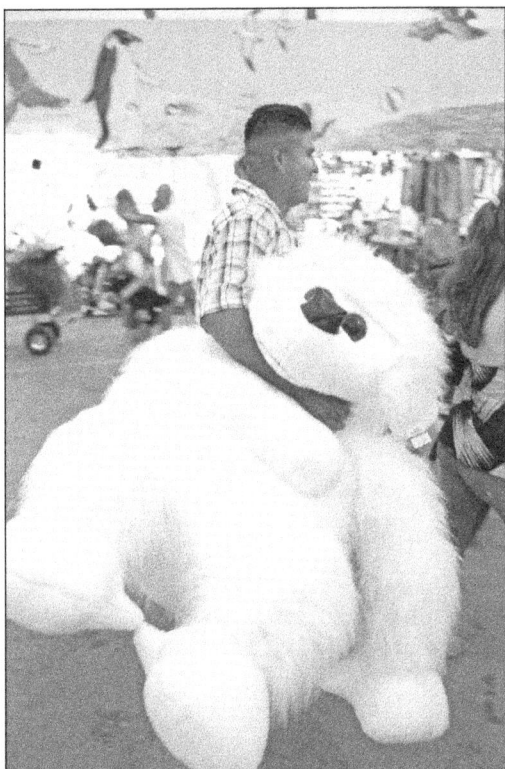

Many love the fair for its entertainment value. Carnival rides and side games are a major attraction for many fairgoers. A big draw for some are the plush toys that may be won by skillfully throwing darts, tossing balls, or shooting hoops with an accurate aim that can't be fooled by often deliberately misaligned targets. (Courtesy 22nd DAA.)

The high point of the Independence Day events is the fireworks display. Not only is the grandstand packed with spectators for this event, which year after year gains in size an intensity and is skillfully choreographed to music, but neighboring streets in Solana Beach and Del Mar also serve as striking vantage points. (Courtesy 22nd DAA.)

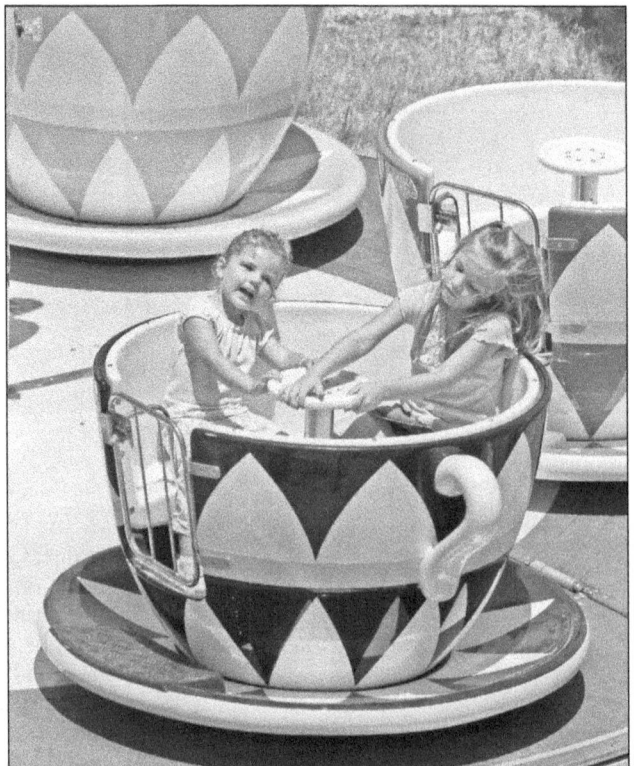

Children have loved the county fair since its inception in 1880, when school was closed and everyone was encouraged to attend National City's first fair. The carnival rides have a separate section for the younger ride seekers. Known as the Infield Kiddie Land, the rides in this children's fun zone don't have many of the height restrictions that are required to experience some of the fairgrounds' more extreme rides. (Courtesy 22nd DAA.)

Today's fair attracts huge crowds during its three-week run in the summer. A far cry from Kimball's first two-day fair held in National City that initiated San Diego County's fair history, shoulder-to-shoulder people wind their way down the Midway, past the famed Del Mar Fairgrounds food vendors, toward the carnival rides. In 2007, there were a recorded 1,265,997 fairgoers in attendance. (Courtesy 22nd DAA.)

Each fair season there is a novel food attraction to add to the appeal of the fair's famed food. In 2007, Pepsi, a major fair sponsor, could be found fried as a fritter. There was also a Krispy Kreme chicken doughnut that year. More traditional "Fair Fare," though, includes the aromatic barbecued beef and chicken pictured and some surprising upscale dining and sipping experiences. (Courtesy 22nd DAA.)

When moving from one side of the fairgrounds to the other, riding by Skylift saves a lot of shoe leather and sore muscles. It also offers a relaxed, spectacular, aerial view of the fair and its surrounding coastal cities, the San Dieguito lowlands, and the Pacific Ocean. (Courtesy 22nd DAA.)

Fairgrounds entertainment lasts well beyond sunset. This nighttime photograph captures the haunting beauty of the well-lit Ferris wheel, a quintessential favorite carnival ride, whose multi-colored lights are reflected in the nearby San Dieguito River. (Courtesy 22nd DAA.)

The Chevrolet Del Mar Arena features exciting motor sports entertainment and animal events, such as monster trucks, an Action Sport Expo, horse shows, camel and ostrich races, and much more. This 2007 photograph has captured the Chino Police Force's monster truck as it ferries willing riders over a dirt-filled obstacle course. (Courtesy 22nd DAA.)

A recent addition to the fairgrounds' lineup is San Diego's largest, scariest, and most haunted attraction: the Scream Zone, a nighttime fright fest designed to spook even the most hardened of patrons. Coinciding with Halloween, the attraction features the Haunted Hayride, the Chamber, and blood-curdling scenes in the House of Horrors. (Courtesy 22nd DAA.)

The month of December sees the Del Mar Fairgrounds stage its Holiday of Lights, a spectacular drive-through light show set to music. Themed for the holiday season, it has more than 400 family-oriented displays that include Candy Cane Lane, Toyland, San Diego County Fair, Treasures by the Lake, Del Mar Racetrack, the Twelve Days of Christmas, and Elves at Play. (Courtesy 22nd DAA.)

Today's fairgrounds has eclipsed the vision of Jim Franks and his 1936 board of directors. A state-of-the-art complex now stands on the site where Franks helped mix the adobe mud for thousands of bricks that year. Located 20 miles north of downtown San Diego, at 2260 Jimmy Durante Boulevard in the city of Del Mar, California, the Del Mar Fairgrounds is a first-class operation. With permanent exhibition halls, a remodeled grandstand, an air-conditioned Mission Tower building, the Equus Skybox, offsite wagering, and modern offices, the site of the Del Mar Fairgrounds remains an unparalleled community resource and a world-class destination. (Courtesy 22nd DAA.)

BIBLIOGRAPHY

Annual Report of the State Board of Horticulture of the State of California for 1889. Sacramento: State Printing, 1890.

BING Magazine. "It's the Season that Bing Crosby Loved Most." County Durham, UK: International Club Crosby, 2007.

Franks, James E. *Del Mar Fair Ground Report.* San Diego: Self-published paper, 1966.

Hanks Ewing, Nancy. *Del Mar Looking Back.* Del Mar, CA: The Del Mar Historical Society, 1991.

Jerstad Cater, Judy. *The Del Mar Fairgrounds, Design, Construction, Early Operation, 1933–1941.* San Diego: Uncirculated USD master's thesis, 1984.

Kimball, Frank. *The Kimball Diaries.* Excerpts from 1880. National City, CA: Unpublished.

Macfarlane, Malcolm. *Bing Crosby: Day by Day.* Lanham, MD: Scarecrow Press, 2001.

Murray, William. *Del Mar, Its Life and Good Times.* Del Mar, CA: Del Mar Thoroughbred Club, 2003.

The National Horse Show Program. Del Mar, CA: 22nd DAA. 1995, 2005.

San Diego County Fair Horse Show Program. Del Mar, CA: 22nd DAA, 1940.

San Diego County Fair Premium Lists. Del Mar, CA: 22nd DAA, 1936–1948.

San Diego Union Newspaper archives. San Diego: Copley Press, 1880–2000.

San Dieguito Citizen Newspaper archives. Solana Beach, CA: The Lapham Family, 1946–1956.

INDEX

ACROSS AMERICA, PEOPLE ARE DISCOVERING
SOMETHING WONDERFUL. *THEIR HERITAGE.*

Arcadia Publishing is the leading local history publisher in the United States. With more than 4,000 titles in print and hundreds of new titles released every year, Arcadia has extensive specialized experience chronicling the history of communities and celebrating America's hidden stories, bringing to life the people, places, and events from the past. To discover the history of other communities across the nation, please visit:

www.arcadiapublishing.com

Customized search tools allow you to find regional history books about the town where you grew up, the cities where your friends and family live, the town where your parents met, or even that retirement spot you've been dreaming about.

www.ingramcontent.com/pod-product-compliance
Lightning Source LLC
Chambersburg PA
CBHW080618110426

42813CB00006B/1543